10 Indiana ILEARN Grade 6 Math Practice Tests

The Ultimate Test Prep Collection with Answer Explanations

Dr. A. Nazari

Copyright © 2026 Dr. A. Nazari

Published by View Math Education

ViewMath.com

10 Practice Tests

The Grand Championship Collection

Welcome, future Math Champion!

*You hold the **ultimate collection** —
ten full-length practice tests designed to take you
from first attempt to **complete mastery.***

- *Conquer every Grade 6 topic*
- *Build unshakeable confidence*
- *Rise from Bronze to Gold to Champion*
- *Arrive at test day fully prepared*

The championship begins now.

★ ★ ★ ★ ★

> **"** *Ten tests may seem like a marathon, but champions are made one step at a time. Trust the process!* **"**

The Champion's Path

Your 4-phase journey from Bronze to Champion

Bronze Round (Tests 1–3)

Your warm-up matches. Take these **untimed** to learn the format and set your baseline. Read the answer explanations after each test — this is where you build your foundation.

Silver Round (Tests 4–6)

Set a timer for **75 minutes**. Focus on the topics that tripped you up in Bronze. Practice showing your work on every problem. Your accuracy should be climbing.

Gold Round (Tests 7–9)

Full timed conditions (**60 minutes**). Simulate the real exam environment. Review only the questions you missed — targeted practice is the key to gold.

Championship Final (Test 10)

Your final match. Full exam conditions — timed, quiet, no breaks. This is your victory lap. Show yourself how far you've come!

Your Championship Kit

- **10 Full-Length Practice Tests** — every Grade 6 topic
- **Formula Reference Sheet**
- **Complete Answer Key** with explanations
- **Championship Scoreboard** to track your rise

Champion's Tip: Space your tests 2–3 days apart. Use the days in between for targeted review. By Test 10, you'll be amazed at your transformation.

👑 The Champion's Playbook 👑

I   **Read every question twice.** The first read tells you the topic. The second tells you exactly what to solve for. Champions never skim.

II   **Mark the clues.** Circle key numbers, underline the question, and cross out information that's just there to distract you.

III   **Choose your strategy.** Before touching pencil to paper, decide: Am I setting up a ratio? Solving an equation? Finding area? Name the approach.

IV   **Solve, then match.** For multiple choice — work the problem on scratch paper first, then find your answer among the choices.

V   **Eliminate and conquer.** Cross out obviously wrong answers. If you're left with two, you've already doubled your odds. Make an educated pick.

VI   **Estimate to verify.** After solving, ask: "Is this answer reasonable?" A quick mental estimate catches most calculation errors.

VII   **Leave nothing blank.** Even a well-reasoned guess is worth more than an empty space. Use partial work to support your answer.

🕐 Timing Mastery

Tests 1–3: **Untimed** (build foundation) › Tests 4–6: **75 min** (build speed) › Tests 7–10: **60 min** (championship conditions)

⭐ Grade 6 Championship Topics

🏅 Ratios & Proportions   🏅 Integers & Rational Numbers   🏅 Expressions & Equations

🏅 Geometry & Measurement   🏅 Statistics & Data Analysis

*A true champion isn't someone who never makes mistakes — it's someone who learns from **every single one**. After each test, review your errors carefully. That's where the real growth happens.*

Find more at
ViewMath.com/IN-Grade6

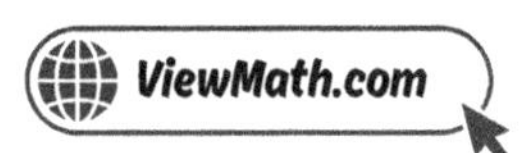

 The Champion's Toolkit

Prepare your workspace before each championship round

🏆 Required Equipment

✏️	**Sharpened Pencils**	Two #2 pencils — champions always have a backup
◢	**Quality Eraser**	A clean, soft eraser that won't smudge your work
📄	**Scratch Paper**	Blank paper for calculations, diagrams, and number lines
📏	**Ruler**	Essential for geometry and coordinate plane questions
⏱️	**Timer**	Begin using from the Silver Round onward
🔊	**Quiet Workspace**	A calm, well-lit area free from distractions

🏅 Permitted in Competition

- ✔ Pencil and eraser
- ✔ Scratch paper (provided)
- ✔ Ruler (if specified)
- ✔ Formula reference in this book

🚫 Not Permitted

- ✖ Calculators
- ✖ Electronic devices
- ✖ Textbooks or notes
- ✖ Outside help

- *With 10 tests, space them **2–3 days apart**. This gives time to review mistakes and study between rounds.*

- *Let your child take Tests 1–3 untimed to build familiarity and establish a baseline.*

- *After each test, go through the Answer Key together. Focus on **understanding the reasoning**, not memorizing answers.*

- *Use the Championship Scoreboard to visualize long-term progress. Celebrate improvements at every tier!*

- *Pair with our **Grade 6 Math Study Guide** for topics that need sustained attention.*

Find more at
ViewMath.com/IN-Grade6

Formula Reference Sheet

🔷 Area Formulas

Rectangle	$A = l \times w$
Parallelogram	$A = b \times h$
Triangle	$A = \dfrac{1}{2} \times b \times h$
Trapezoid	$A = \dfrac{1}{2}(b_1 + b_2) \times h$

🧊 Volume

Rectangular Prism $\quad V = l \times w \times h$

🧊 Surface Area

Find the area of each face, then add them all up.

Rectangular Prism:

$SA = 2lw + 2lh + 2wh$

⬇️ Order of Operations

P Parentheses first

E Exponents

M/D Multiply & Divide (left to right)

A/S Add & Subtract (left to right)

% Ratios & Percents

Ratio: $a : b$ or $\dfrac{a}{b}$

Unit rate: amount per 1 unit

Percent: a ratio out of 100

$Part = Percent \times Whole$

⚖️ Integers & Absolute Value

Integers:

$\dots, -3, -2, -1, 0, 1, 2, 3, \dots$

$|-5| = 5 \quad |5| = 5$

Absolute value = distance from 0

X^1 Expressions & Equations

Exponent: $3^4 = 3 \times 3 \times 3 \times 3 = 81$

Variable: a letter that stands for a number

Equation: two expressions joined by $=$

Inequality: uses $<, >, \leq, \geq$

⊕ Coordinate Plane

Ordered pair: (x, y)

x-axis: horizontal y-axis: vertical

Origin: $(0, 0)$

Four quadrants (I, II, III, IV)

📊 Statistics

Mean: sum of values $\div$ count

Median: middle value (sorted)

Range: max $-$ min

Championship Scoreboard

Track your rise through every championship round

Champion's Name: ___________________________

Round	Tier	Date	Score	Rating
1	Bronze		/	
2	Bronze		/	
3	Bronze		/	
4	Silver		/	
5	Silver		/	
6	Silver		/	
7	Gold		/	
8	Gold		/	
9	Gold		/	
10	♔		/	

My strongest topics (where I consistently score well):

____

____

Topics I improved on the most from Bronze to Gold:

____

____

My score trend (Bronze avg → Gold avg → Championship):

____

____

One strategy that helped me improve the most:

____

____

My confidence level for the real test (1–10): _________ / 10

Find more at
ViewMath.com/IN-Grade6

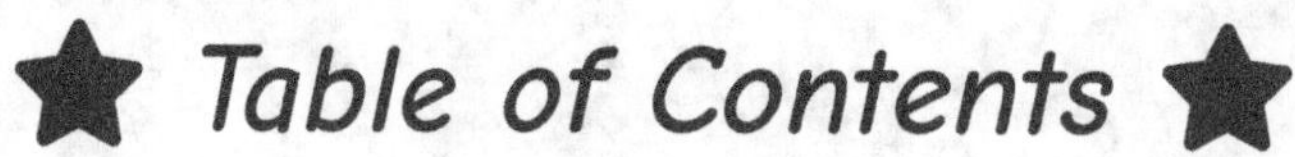

★ Table of Contents ★

Here's what we'll explore together!

 Let's learn and have fun!

Practice Test 1

 30 Questions

✏️ Before You Start ✏️

- ✔ **Read each question carefully** before choosing your answer.
- ✔ **Show your work** on scratch paper when you need to.
- ✔ **Skip hard questions** and come back to them later.
- ✔ **Check your answers** when you're done.
- ✔ **Take your time** — there's no rush!

⭐ You've Got This! ⭐

Do your best and show what you know!

1. A garden has flowers and vegetables in a ratio of 4 : 1. There are 20 flowers. How many vegetables are there?

Your Answer:

2. The graph below shows the distance traveled by a delivery truck over time.

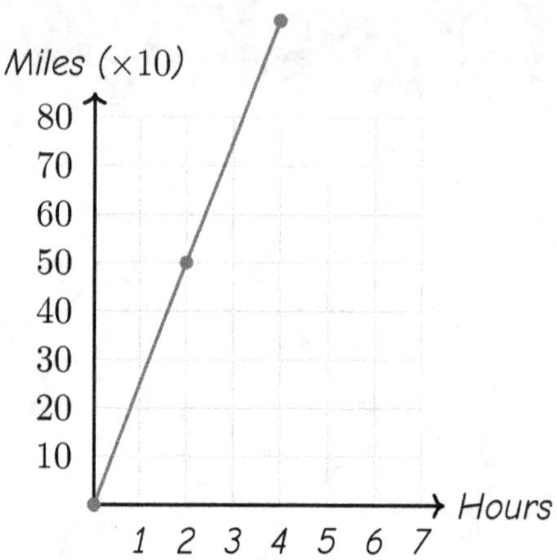

What is the truck's unit rate in miles per hour?

(A) 50 miles per hour

(B) 20 miles per hour

(C) 25 miles per hour

(D) 10 miles per hour

3. The table and partial graph below show a ratio relationship.

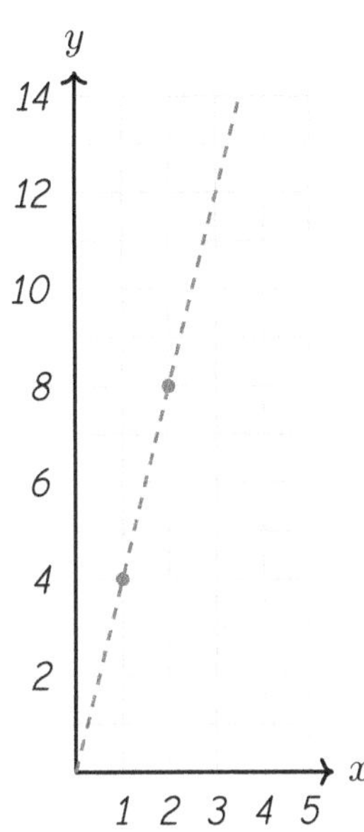

x	y
1	4
2	8
3	?

What is the missing value of y when $x = 3$?

(A) 10

(B) 12

(C) 14

(D) 16

4. What is 72% written as a decimal?

(A) 72.0

(B) 7.2

(C) 0.072

(D) 0.72

5. The bar model below compares different length units.

1 yard		
1 ft	1 ft	1 ft

If a rope is 7 yards long, how many feet is it?

(A) 10

(B) 14

(C) 21

(D) 28

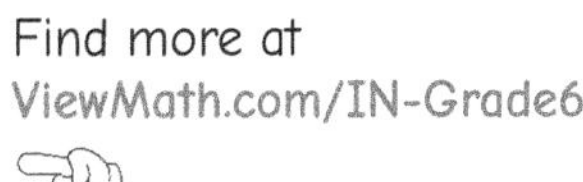
Find more at
ViewMath.com/IN-Grade6

6. Compute $9{,}072 \div 42$.

Your Answer:

7. The number line below shows two points, A and B.

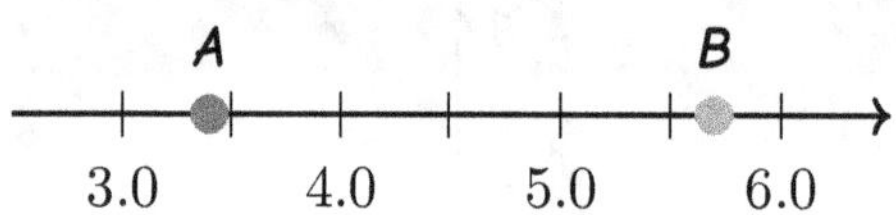

What is the distance from A to B?

(A) 2.3

(B) 2.1

(C) 9.1

(D) 3.3

8. What is the LCM of 4 and 10?

(A) 20

(B) 2

(C) 10

(D) 40

9. Convert $-\dfrac{5}{8}$ to a decimal and state which two consecutive integers it falls between on a number line.

Your Answer:

10. On a town map, the park is at $(-6, 4)$ and the museum is at $(2, 4)$. Each grid unit represents one block. How many blocks apart are the park and the museum? Show your work.

Your Answer:

Find more at
ViewMath.com/IN-Grade6

11. The diagram shows two groups. Which expression represents the total?

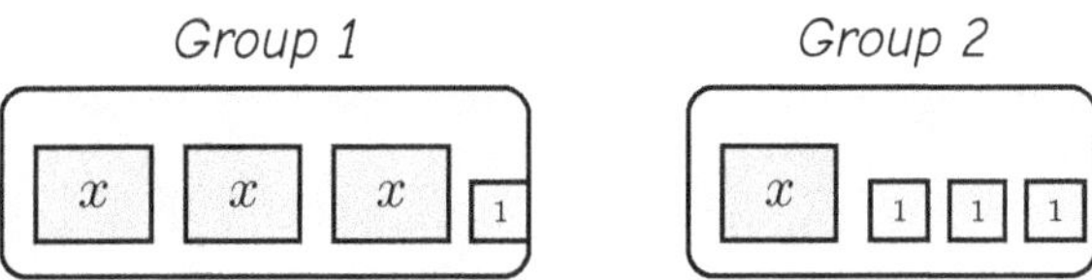

(A) $3x + 4$

(B) $4x + 4$

(C) $4x + 3$

(D) $3x + 1 + x + 3$

12. The bar diagram below shows that Jake's age plus 4 years equals his sister's age. His sister is 13 years old. Write and solve an equation for Jake's age j.

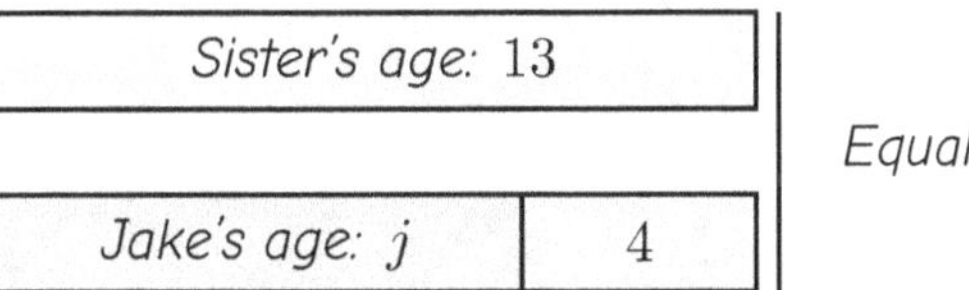

Your Answer.

13. Write an inequality: The pool is open when the temperature is above 75°F. Let t = temperature.

Your Answer.

14. When graphing $x > 4$ on a number line, what kind of circle goes at 4?

(A) Closed circle (filled in)

(B) Open circle (not filled in)

(C) No circle — just an arrow

(D) A square

15. A school bus holds 50 students. How many buses b are needed for s students? Write an equation.

Your Answer:

16. A triangle has base 14 m and height 8 m. What is its area?

(A) $112\ m^2$

(B) $22\ m^2$

(C) $44\ m^2$

(D) $56\ m^2$

17. A trapezoid has an area of $84\ cm^2$ and bases of 10 cm and 14 cm. What is the height?

Your Answer:

18. A rectangular prism has volume $180\ cm^3$, length 9 cm, and width 5 cm. What is the height?

Your Answer:

19. If a triangle is reflected across the x-axis, what stays the same?

(A) The y-coordinates of all vertices

(B) The size and shape of the triangle

(C) The direction the triangle faces

(D) The sign of the y-coordinates

20. A right triangle has vertices $(1,1)$, $(1,9)$, and $(5,1)$. What is the area?

(A) 32 square units

(B) 16 square units

(C) 20 square units

(D) 8 square units

Find more at
ViewMath.com/IN-Grade6

21. *What is the reflection of $(-3, -5)$ across the x-axis?*

   (A) $(3, -5)$

   (B) $(-3, 5)$

   (C) $(3, 5)$

   (D) $(-5, -3)$

22. *A circular pizza has a diameter of 16 inches. What is the area of the pizza? Use $\pi \approx 3.14$.*

   (A) $50.24 \ in^2$

   (B) $200.96 \ in^2$

   (C) $401.92 \ in^2$

   (D) $803.84 \ in^2$

23. *Maria asks, "How tall are the students in my class?" Why is this a statistical question?*

   (A) Because it asks about height.

   (B) Because different students have different heights.

   (C) Because it mentions a class.

   (D) Because the answer is a number.

24. *Data: $1, 2, 2, 3, 3, 3, 4, 4, 5$. The data has a peak at 3. What does "peak" mean?*

   (A) 3 is the largest value.

   (B) 3 is the value that appears most often.

   (C) 3 is the range.

   (D) 3 is the spread.

25. *Data: $5, 10, 15, 20, 25, 30, 35$. What is the range?*

   (A) 5

   (B) 15

   (C) 20

   (D) 30

Find more at
ViewMath.com/IN-Grade6

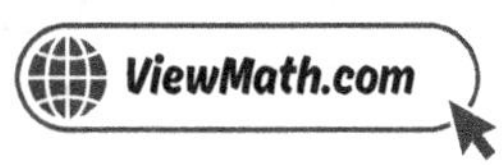

26. *A dot plot shows hours of TV watched daily: 0 (3 dots), 1 (6 dots), 2 (5 dots), 3 (4 dots), 4 (1 dot), 8 (1 dot).*

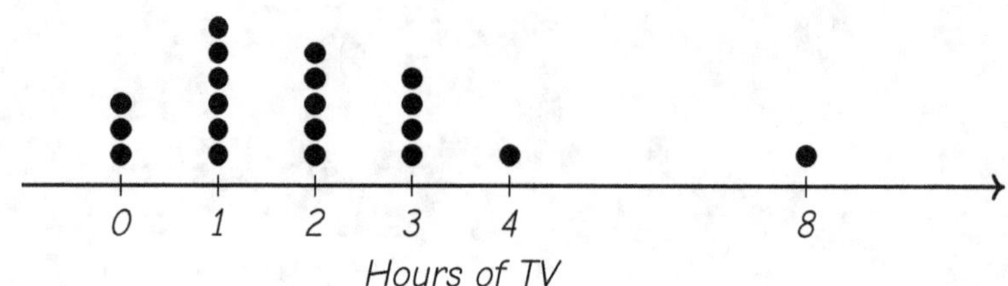

What is the mode, and which value appears to be an outlier?

(A) *Mode = 1 hour, outlier = 0*

(B) *Mode = 1 hour, outlier = 8 hours*

(C) *Mode = 2 hours; outlier = 8 hours*

(D) *Mode = 3 hours; outlier = 4 hours*

27. *Which part of the box plot represents the lowest 25% of the data?*

(A) *The box*

(B) *The left whisker (from min to Q1)*

(C) *The right whisker (from Q3 to max)*

(D) *The line at the median*

28. *A student says "The data set with the larger range is always more spread out than the one with the smaller range." Is this always true? Explain.*

Your Answer:

29. *A standard number cube is rolled. What is the probability of rolling a number less than or equal to 3?*

(A) $\dfrac{1}{3}$

(B) $\dfrac{3}{6}$

(C) $\dfrac{2}{6}$

(D) $\dfrac{4}{6}$

30. *A line graph shows a town's average temperature each month from January (28°F) to May (62°F). The values increase each month. Is this graph showing a positive trend, a negative trend, or no trend?*

Your Answer:

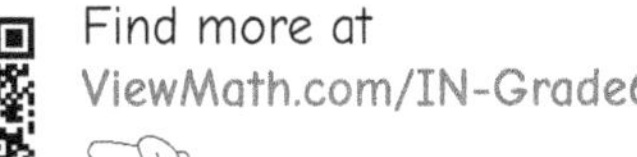
Find more at
ViewMath.com/IN-Grade6

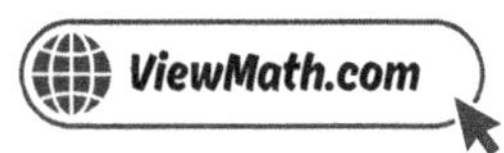

 # End of Practice Test 1 

Great job finishing the test!

 My Score

I got ____________ out of 30 questions right.

*Check your answers in the **Answer Key** at the back of the book.*

💡 *Review any questions you missed. That's how we learn!*

 Check Your Score Online!

Visit **ViewMath Academy** to enter your answers and see which topics you need to review. You can also explore lessons, take quizzes, track your scores, and save your progress!

viewmath.com/score/6.1.IN.16

Or go to viewmath.com/score and enter code: 6.1.IN.16

2

Practice Test 2

✅ 30 Questions

✏️ Before You Start ✏️

- ✓ **Read each question carefully** before choosing your answer.
- ✓ **Show your work** on scratch paper when you need to.
- ✓ **Skip hard questions** and come back to them later.
- ✓ **Check your answers** when you're done.
- ✓ **Take your time** — there's no rush!

⭐ You've Got This! ⭐

Do your best and show what you know!

1. Look at the tape diagram below.

Boys: 

Girls:

There are 24 girls. How many boys are there?

(A) 4

(B) 12

(C) 16

(D) 20

2. The table shows the earnings of two babysitters.

	Hours Worked	Earnings
Lily	5	$60
Noah	4	$52

Part A: Find each babysitter's unit rate (dollars per hour).

Part B: Who earns more per hour? How much more?

Your Answer:

3. Two runners' distances are graphed below.

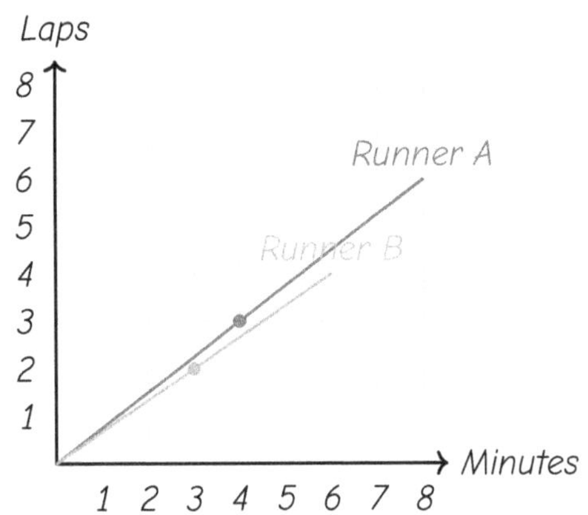

Part A: What is each runner's ratio of laps to minutes?

Part B: Who runs faster? Explain using the graph.

Your Answer

4. Which of the following is equal to 50%?

(A) $\dfrac{1}{5}$

(B) 0.05

(C) $\dfrac{1}{2}$

(D) 5.0

5. A bag weighs 500 grams. What is its weight in kilograms?

(A) 50

(B) 5

(C) 0.5

(D) 0.05

6. An auditorium has 1,890 seats arranged in equal rows of 15. How many rows are there?

(A) 126

(B) 116

(C) 136

(D) 12 R6

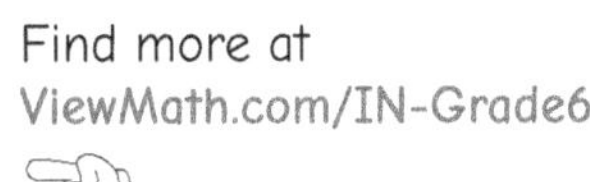
Find more at
ViewMath.com/IN-Grade6

7. *What is* $20.3 - 8.57$?

(A) 12.27

(B) 11.53

(C) 11.73

(D) 28.87

8. *Find both the GCF and the LCM of 8 and 12.*

Your Answer:

9. *Which list shows the numbers in order from least to greatest?*

(A) $-0.5, \ -1.2, \ 0.3, \ 1.5$

(B) $-1.2, \ -0.5, \ 0.3, \ 1.5$

(C) $0.3, \ -0.5, \ 1.5, \ -1.2$

(D) $1.5, \ 0.3, \ -0.5, \ -1.2$

10. *The points $(5, -1)$ and $(-3, -1)$ lie on a horizontal line. Explain step by step how to find the distance between them, and state why the answer is positive.*

Your Answer:

11. *Simplify:* $6 + 2(x + 4)$

(A) $2x + 10$

(B) $8x + 4$

(C) $2x + 14$

(D) $16x$

12. *Write and solve an equation: A number decreased by 19 is 31.*

Your Answer:

13. The table below shows whether each value is a solution to an unknown inequality. Determine the inequality.

Value	Solution?
$x = 2$	No
$x = 3$	No
$x = 4$	No
$x = 5$	Yes
$x = 6$	Yes
$x = 10$	Yes

Your Answer

14. Pick two values that ARE solutions to $x > -1$ and one value that is NOT.

Your Answer

15. In a table of values for $y = 2x + 1$, if $x = 4$, what is y?

(A) 7

(B) 8

(C) 9

(D) 10

16. A triangular garden has a base of 12 ft and a height of 9 ft. How many square feet of soil are needed to cover it?

(A) 108 ft^2

(B) 54 ft^2

(C) 21 ft^2

(D) 42 ft^2

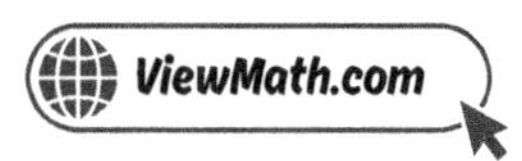

17. A parallelogram has an area of 54 cm^2 and a base of 9 cm. What is the height?

(A) 6 cm

(B) 45 cm

(C) 63 cm

(D) 4.5 cm

18. A shipping box has dimensions 12 in by 8 in by 6 in. How many 1-inch cubes would fit inside?

(A) 26 cubes

(B) 96 cubes

(C) 576 cubes

(D) 288 cubes

19. Reflect the point $(4, -9)$ across the x-axis. What are the new coordinates?

Your Answer:

20. A right triangle has vertices $(2, 1)$, $(2, 7)$, and $(8, 1)$. What is the area?

(A) 36 square units

(B) 18 square units

(C) 12 square units

(D) 24 square units

21. Point $K(-5, -2)$ is translated 7 units right and 4 units up. What are the coordinates of K'?

(A) $(2, 2)$

(B) $(-12, -6)$

(C) $(2, -6)$

(D) $(-12, 2)$

22. Find the area of a semicircle with a diameter of 12 cm. Use $\pi \approx 3.14$.

Your Answer:

Find more at
ViewMath.com/IN-Grade6

23. Which of the following is a statistical question?

  (A) What year was the school built?   (B) What is the name of our principal?

  (C) How many minutes does each student exercise   (D) How many continents are there?
per day?

24. Data: $20, 22, 22, 23, 23, 23, 24, 24, 25$. Describe the center, spread, and shape.

Your Answer:

25. A number line below shows the positions of Q1, the median, and Q3 for a data set.

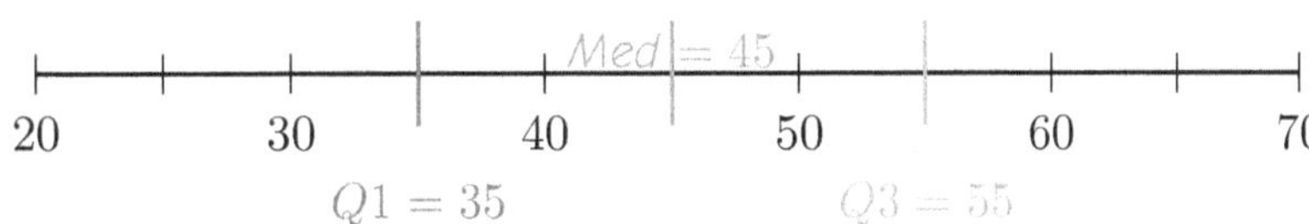

What is the IQR?

  (A) 10   (B) 15

  (C) 20   (D) 35

26. The frequency table below shows the results of a school survey about how students get to school.

Transportation	Frequency
Bus	45
Car	30
Walk	15
Bike	10

Find the total number of students surveyed, the mode, and the percent who walk to school.

Your Answer:

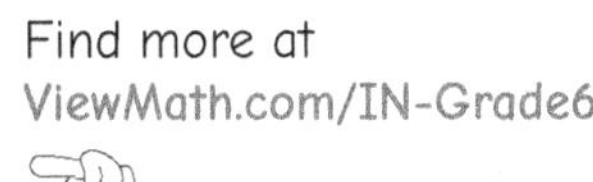

27. Which five values make up the five-number summary?

   (A) Mean, median, mode, range, IQR

   (B) Minimum, Q1, median, Q3, maximum

   (C) Q1, Q2, Q3, Q4, Q5

   (D) Mean, Q1, Q3, range, MAD

28. A coach records sprint times (seconds) for two athletes over 10 races. Athlete A: median $= 12.5$, range $= 1.2$. Athlete B: median $= 12.5$, range $= 3.0$. Which athlete is more consistent?

   (A) Athlete A

   (B) Athlete B

   (C) They are equally consistent.

   (D) Cannot be determined without the IQR.

29. The probability scale below shows four events. Which event has a probability closest to 0.75?

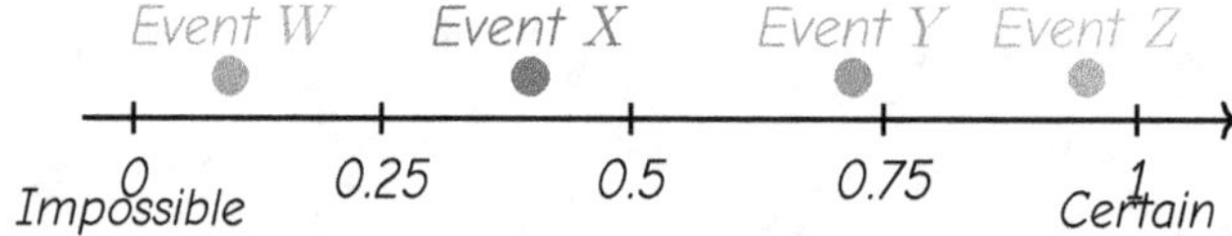

   (A) Event W

   (B) Event X

   (C) Event Y

   (D) Event Z

30. The double bar graph below shows the number of books read by boys and girls in two classes. Use the graph to answer the question.

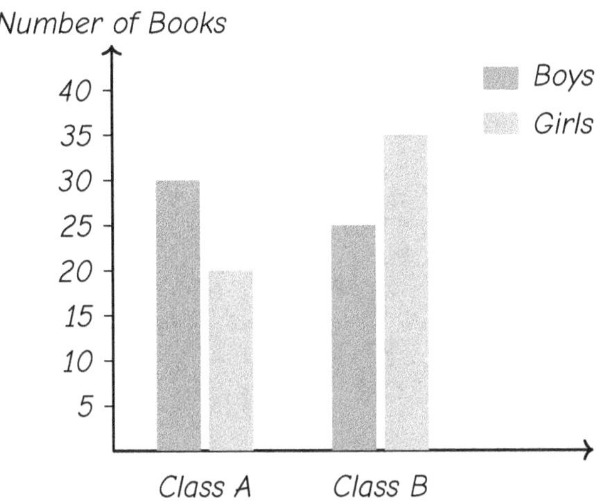

How many more books did boys in Class A read than girls in Class A?

(A) 5

(B) 10

(C) 15

(D) 20

 # End of Practice Test 2 

Great job finishing the test!

 My Score

I got _____________ out of 30 questions right.

*Check your answers in the **Answer Key** at the back of the book.*

💡 *Review any questions you missed. That's how we learn!*

📊 Check Your Score Online!

Visit **ViewMath Academy** to enter your answers and see which topics you need to review. You can also explore lessons, take quizzes, track your scores, and save your progress!

viewmath.com/score/6.1.IN.17

*Or go to **viewmath.com/score** and enter code: 6.1.IN.17*

Practice Test 3

 30 Questions

 Before You Start

- ✓ **Read each question carefully** before choosing your answer.
- ✓ **Show your work** on scratch paper when you need to.
- ✓ **Skip hard questions** and come back to them later.
- ✓ **Check your answers** when you're done.
- ✓ **Take your time** — there's no rush!

 You've Got This! 

Do your best and show what you know!

1. A baker uses 2 eggs **per** 3 cups of flour. Which ratio represents flour to eggs?

(A) $2 : 3$ (B) $3 : 5$

(C) $3 : 2$ (D) $2 : 5$

2. A box of 12 muffins costs \$9. What is the cost per muffin?

(A) \$1.25   (B) \$0.75

(C) \$1.33   (D) \$3.00

3. A ratio graph passes through $(6, 2)$. What is the value of y when $x = 15$?

(A) 3   (B) 5

(C) 7   (D) 10

4. A store has 200 items. 50% are on sale. How many items are on sale?

(A) 50   (B) 150

(C) 100   (D) 200

5. Convert 156 inches to feet.

(A) 11   (B) 12

(C) 13   (D) 14

6. What is $756 \div 3$?

(A) 252   (B) 250

(C) 258   (D) 202

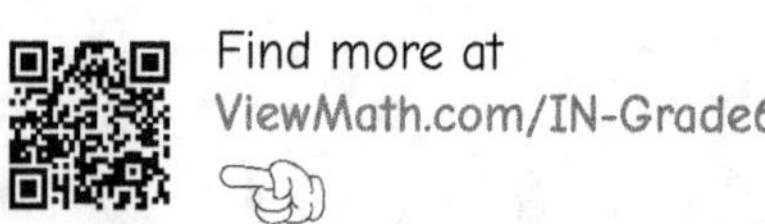

Find more at
ViewMath.com/IN-Grade6

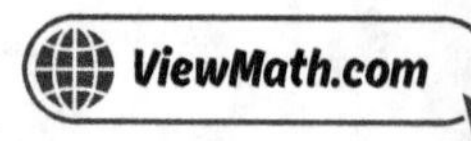

7. *What is* $0.12 \times 0.5?$

   (A) 0.6

   (B) 0.06

   (C) 0.006

   (D) 6.0

8. *What is the GCF of 36 and 48?*

   (A) 12

   (B) 6

   (C) 24

   (D) 144

9. *Between which two consecutive integers is* $\dfrac{11}{4}$ *located on a number line?*

   (A) 1 and 2

   (B) 2 and 3

   (C) 3 and 4

   (D) 0 and 1

10. *Which pair of points has the greatest distance between them?*

    (A) $(1, 3)$ *and* $(1, 7)$

    (B) $(-2, 5)$ *and* $(4, 5)$

    (C) $(0, -3)$ *and* $(0, 2)$

    (D) $(-1, 4)$ *and* $(3, 4)$

11. *Simplify:* $4(3y - 2)$

    (A) $12y - 2$

    (B) $12y - 8$

    (C) $7y - 2$

    (D) $7y - 6$

12. *Solve:* $15a = 105$

    Your Answer:

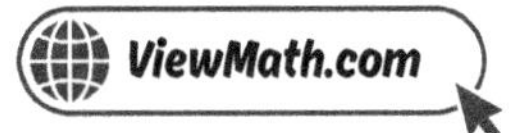

13. Which inequality represents "a number y is fewer than 20"?

(A) $y > 20$

(B) $y \geq 20$

(C) $y < 20$

(D) $y \leq 20$

14. Write the inequality that matches this description: closed circle at -4, shade to the left.

Your Answer:

15. The number of inches i equals 12 times the number of feet f: $i = 12f$. How many inches are in 5 feet?

(A) 17 inches

(B) 48 inches

(C) 60 inches

(D) 125 inches

16. A triangle has base 15 cm and height 4 cm. Sam says the area is 60 cm^2. What mistake did Sam make?

(A) He used the wrong formula entirely.

(B) He forgot to divide by 2.

(C) He added instead of multiplied.

(D) He subtracted the height from the base.

17. *What is the area of the trapezoid shown below?*

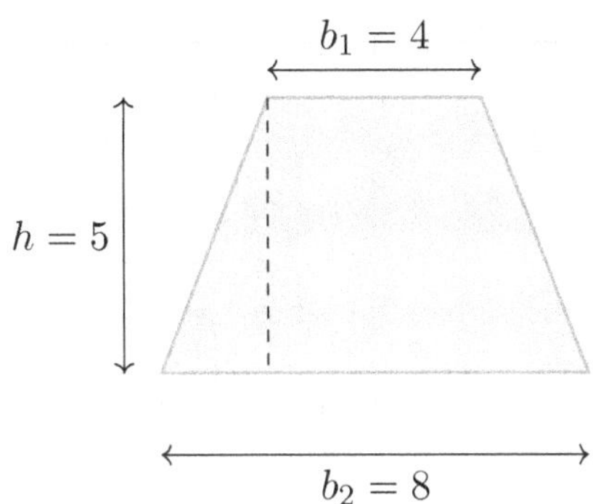

(A) 20 *square units*

(B) 40 *square units*

(C) 60 *square units*

(D) 30 *square units*

18. *A rectangular prism has length $\frac{1}{2}$ m, width $\frac{1}{2}$ m, and height $\frac{1}{2}$ m. What is the volume?*

(A) $\frac{1}{8}\ m^3$

(B) $\frac{1}{4}\ m^3$

(C) $\frac{3}{2}\ m^3$

(D) $\frac{1}{2}\ m^3$

19. *A rectangle has vertices $(-4, 5)$, $(6, 5)$, $(6, -3)$, $(-4, -3)$. What is the perimeter?*

(A) 18 *units*

(B) 26 *units*

(C) 36 *units*

(D) 80 *units*

20. *An L-shaped figure has vertices $(0, 0)$, $(8, 0)$, $(8, 3)$, $(4, 3)$, $(4, 7)$, and $(0, 7)$. A student splits it into two rectangles. Which pair of rectangles correctly covers the figure?*

(A) 8×7 *and* 4×3

(B) 8×3 *and* 4×4

(C) 4×7 *and* 4×3

(D) 8×3 *and* 4×7

21. A triangle has vertices $D(1,4)$, $E(5,4)$, and $F(3,7)$. The triangle is reflected across the y-axis. What are the coordinates of F'?

(A) $(3,-7)$

(B) $(-3,-7)$

(C) $(-3,7)$

(D) $(7,-3)$

22. The circle below has a diameter of $14\ cm$. Find both the circumference and the area of the circle. Use $\pi \approx 3.14$.

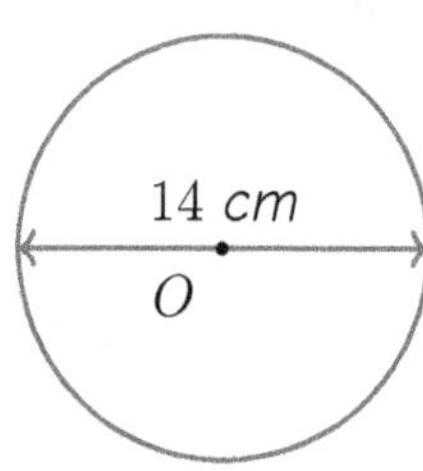

Your Answer:

23. A librarian asks, "How many books were checked out today?" Is this a statistical question?

(A) Yes, because it is about books.

(B) Yes, if she plans to ask the same question over many days.

(C) No, because on one specific day there is only one answer.

(D) No, because libraries always have the same number of checkouts.

24. Data: $15, 16, 16, 17, 17, 17, 18, 100$. How does the outlier 100 affect the center?

(A) It pulls the center toward 100, making it higher than typical values.

(B) It has no effect on the center.

(C) It makes the center equal to 100.

(D) It pulls the center lower.

25. Which statement about MAD is true?

(A) MAD measures the middle value of the data.

(B) MAD tells you the average distance of each value from the mean.

(C) MAD is always larger than the range.

(D) MAD equals the range divided by 2.

26. A dot plot of daily steps: 3000 (1), 4000 (2), 5000 (6), 6000 (4), 7000 (2). What is the mode? What is the range?

27. Data: $1, 3, 5, 7, 9, 11, 13, 15, 17, 19$. What is the IQR?

(A) 8

(B) 10

(C) 14

(D) 18

28. A data set has a range of 40 but an IQR of only 8. What does this tell you?

(A) All values are close together.

(B) Most values are clustered, but there are extreme values far from the center.

(C) The median equals the mean.

(D) The data is symmetric.

29. A standard number cube (die) is rolled once. What is the probability of rolling a 4?

(A) $\dfrac{4}{6}$

(B) $\dfrac{1}{4}$

(C) $\dfrac{1}{6}$

(D) $\dfrac{1}{3}$

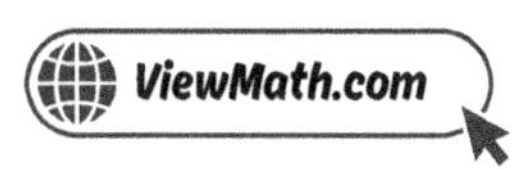

30. *A frequency table records the number of pets owned by 25 students. The frequencies are:* 0 pets (6), 1 pet (10), 2 pets (5), 3 pets (4). *How many students own at least 2 pets?*

(A) 5

(B) 9

(C) 4

(D) 15

 # End of Practice Test 3 

Great job finishing the test!

 My Score

I got _________ out of 30 questions right.

Check your answers in the **Answer Key** at the back of the book.

Review any questions you missed. That's how we learn!

📊 Check Your Score Online!

Visit **ViewMath Academy** to enter your answers and see which topics you need to review. You can also explore lessons, take quizzes, track your scores, and save your progress!

viewmath.com/score/6.1.IN.18

Or go to viewmath.com/score and enter code: 6.1.IN.18

Practice Test 4

 30 Questions

✏️ Before You Start ✏️

✔ **Read each question carefully** before choosing your answer.

✔ **Show your work** on scratch paper when you need to.

✔ **Skip hard questions** and come back to them later.

✔ **Check your answers** when you're done.

✔ **Take your time** — there's no rush!

 ⭐ You've Got This! ⭐ 

Do your best and show what you know!

1. *A store sells* 3 *pencils **for every*** 1 *eraser. Which ratio represents pencils to erasers?*

   (A) 1 : 3   (B) 3 : 1

   (C) 3 : 4   (D) 1 : 4

2. *A bike travels at a unit rate of* 14 *miles per hour. How long will it take to travel* 49 *miles?*

3. *A line passes through the origin and the point* $(5, 15)$. *What is the ratio* x *to* y?

   (A) 1 : 5   (B) 5 : 1

   (C) 1 : 3   (D) 3 : 1

4. *Convert* 0.875 *to a percent and then to a fraction in simplest form.*

5. *How many inches are in* 5 *feet?*

   (A) 50   (B) 55

   (C) 60   (D) 72

6. *What is* $1,575 \div 5$?

   (A) 305   (B) 3,150

   (C) 315   (D) 351

Find more at
ViewMath.com/IN-Grade6

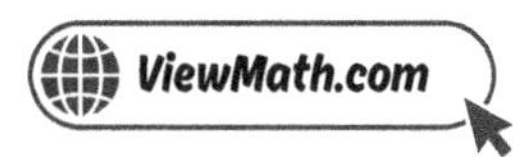

7. *When you multiply 0.6×0.7, how many decimal places should the product have?*

(A) 0   (B) 1

(C) 2   (D) 3

8. *A teacher has 48 pencils and 80 erasers to put into identical prize bags with no supplies left over. What is the greatest number of bags she can make?*

Your Answer:

9. *On a number line, where is $\dfrac{5}{8}$ located?*

(A) Between 0 and 1   (B) Between 1 and 2

(C) Between -1 and 0   (D) Between 2 and 3

10. *What is the distance between $(2, -3)$ and $(2, 5)$?*

(A) -8 (B) 2

(C) 4   (D) 8

11. *A student says $2x + 3x = 5x^2$. Is this correct?*

(A) Yes   (B) No — the answer is $6x^2$

(C) No — the answer is $5x$ (D) No — the answer is $6x$

12. *Solve: $\dfrac{w}{8} = 9$*

Your Answer:

Find more at
ViewMath.com/IN-Grade6

13. The table below matches phrases to inequality symbols. Which row has an error?

Row	Phrase	Symbol
1	At least	$\geq$
2	No more than	$\leq$
3	Fewer than	$<$
4	At most	$<$

(A) Row 1

(B) Row 2

(C) Row 3

(D) Row 4

14. A student says the graph of $x < 8$ and the graph of $x \leq 8$ look exactly the same. Is this correct?

15. A car wash charges $10 per car. Which variable goes on the x-axis when graphing?

(A) Total cost

(B) Number of cars

(C) Price per car

(D) Profit

16. Look at the two triangles on the grid. Which triangle has a greater area?

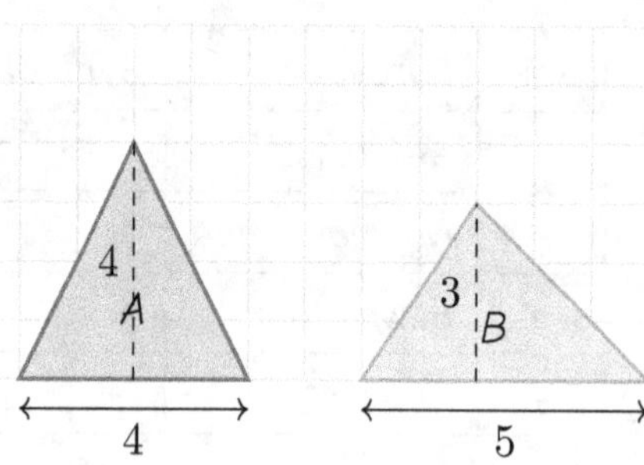

  (A) Triangle A   (B) Triangle B

  (C) They have the same area.   (D) Not enough information to tell.

17. A trapezoid has bases $b_1 = 4$ cm and $b_2 = 4$ cm with height 6 cm. What shape is this trapezoid actually equivalent to?

  (A) A triangle   (B) A parallelogram

  (C) A circle   (D) A pentagon

18. A rectangular prism has edges 1.5 cm, 2 cm, and 4 cm. What is the volume?

  (A) $12 \ cm^3$ (B) $7.5 \ cm^3$

  (C) $8 \ cm^3$ (D) $6 \ cm^3$

19. Points $(1, 4)$ and $(1, -2)$ form a vertical segment. What is the length?

  (A) 2 units   (B) 4 units

  (C) 5 units   (D) 6 units

20. *A rectangle has vertices* $(-1, 3)$, $(5, 3)$, $(5, -2)$, *and* $(-1, -2)$. *What is the area?*

(A) 25 *square units*   (B) 30 *square units*

(C) 22 *square units*   (D) 35 *square units*

21. *A point at* $(2, -6)$ *is reflected across the y-axis. Where does it land?*

(A) $(2, 6)$ (B) $(-2, 6)$

(C) $(-2, -6)$ (D) $(6, -2)$

22. *Circle A has a radius of 3 cm and Circle B has a radius of 6 cm. How many times larger is the area of Circle B than the area of Circle A?*

(A) 2 *times*   (B) 3 *times*

(C) 4 *times*   (D) 6 *times*

23. *A student surveyed classmates about how many books they read last month and made the dot plot below.*

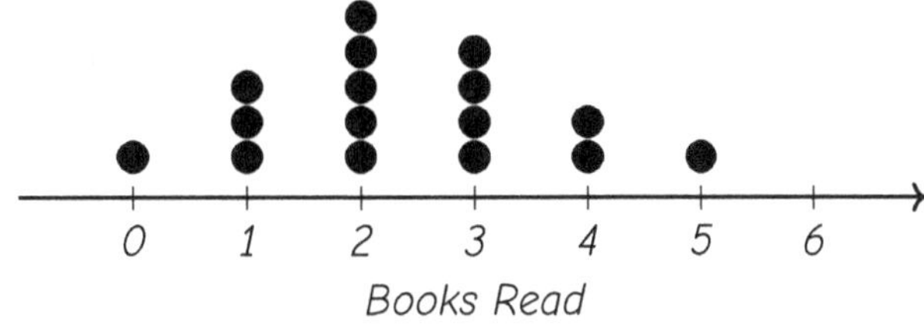

What does this dot plot confirm about the survey question "How many books did you read last month?"

(A) *It is not statistical because only whole numbers appear.*

(B) *It is statistical because the answers vary from 0 to 5.*

(C) *It is not statistical because some students read the same number.*

(D) *It is statistical because there are exactly 16 dots.*

24. *Describe the shape of the data set:* $1, 2, 3, 3, 4, 4, 4, 5, 5, 5, 5.$

Your Answer:

25. *Data (in order):* $4, 6, 8, 10, 12, 14, 16.$ *What is Q3?*

(A) 10

(B) 12

(C) 14

(D) 16

26. *A dot plot of temperatures shows:* $68°F$ *(1 dot),* $70°F$ *(3 dots),* $72°F$ *(4 dots),* $74°F$ *(2 dots). What is the range?*

(A) 4

(B) 6

(C) 10

(D) 74

27. *If the whiskers of a box plot are both short and the box is narrow, what does this tell you?*

(A) The data is very spread out.

(B) The data is highly consistent with little variability.

(C) There are many outliers.

(D) The data set is very large.

28. *Data:* $12, 15, 18, 20, 22, 25, 28, 30.$ *What is the IQR?*

(A) 10

(B) 11

(C) 18

(D) 8

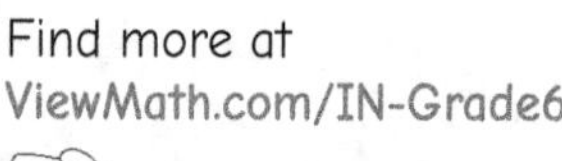

29. Use the probability scale below. Place each event at the correct position by writing its letter on the scale. Then state each probability as a fraction or decimal.

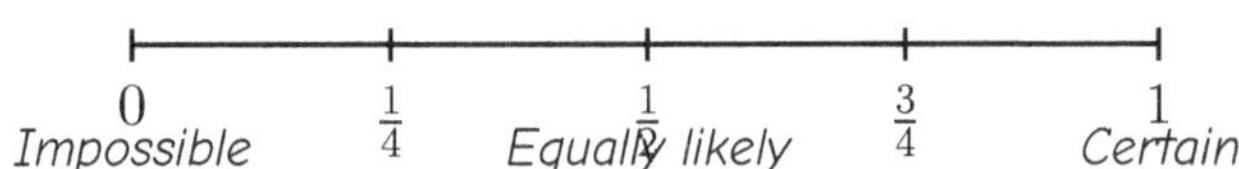

Event A: Rolling a 7 on a standard number cube.

Event B: Flipping tails on a fair coin.

Event C: Drawing a red marble from a bag that has 6 red and 2 blue marbles.

Event D: Rolling a number less than 7 on a standard number cube.

Your Answer

30. On a line graph, the line goes steeply upward from April to May and then stays flat from May to June. What does this tell you?

(A) The value decreased from April to June

(B) The value increased sharply, then stayed the same

(C) The value stayed the same the entire time

(D) The value increased at a steady rate

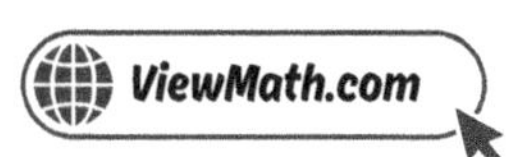

 End of Practice Test 4 

Great job finishing the test!

✅ My Score

I got ___________ out of 30 questions right.

*Check your answers in the **Answer Key** at the back of the book.*

💡 *Review any questions you missed. That's how we learn!*

📊 Check Your Score Online!

Visit **ViewMath Academy** to enter your answers and see which topics you need to review. You can also explore lessons, take quizzes, track your scores, and save your progress!

viewmath.com/score/6.1.IN.19

Or go to *viewmath.com/score* and enter code: *6.1.IN.19*

Practice Test 5

30 Questions

✏️ Before You Start ✏️

- ✓ **Read each question carefully** before choosing your answer.
- ✓ **Show your work** on scratch paper when you need to.
- ✓ **Skip hard questions** and come back to them later.
- ✓ **Check your answers** when you're done.
- ✓ **Take your time** — there's no rush!

⭐ You've Got This! ⭐

Do your best and show what you know!

1. A class votes on two activities: hiking and swimming. The ratio of votes for hiking to swimming is $3 : 2$. There are 25 votes total. How many voted for swimming?

Your Answer:

2. A car travels 240 miles using 8 gallons of gas. What is the unit rate in miles per gallon?

(A) 8

(B) 32

(C) 30

(D) 248

3. A graph of equivalent ratios passes through $(0,0)$ and $(3,2)$. Which statement is true?

(A) The point $(9,4)$ is on the line.

(B) The point $(6,4)$ is on the line.

(C) The point $(6,5)$ is on the line.

(D) The point $(9,8)$ is on the line.

4. The bar model below shows the results of a class vote on a field trip destination.

Zoo (60%)	Farm	Park

Total: 100%

If the Farm section is twice as wide as the Park section, what percent voted for the Farm?

(A) 10%

(B) 40%

(C) $\dfrac{20}{3}\%$

(D) $\approx 27\%$

5. How many meters are in 4.5 kilometers?

(A) 45

(B) 450

(C) 4,500

(D) 45,000

Find more at
ViewMath.com/IN-Grade6

6. *Compute* $4{,}515 \div 15$. *Be careful with every digit of the quotient.*

7. *What is* 1.2×0.3?

(A) 3.6   (B) 0.36

(C) 36   (D) 0.036

8. *You have a ribbon that is 72 inches long and another that is 96 inches long. You want to cut both ribbons into equal-length pieces with no ribbon left over. What is the longest possible piece?*

(A) 24   (B) 8

(C) 288   (D) 12

9. *Between which two consecutive integers is* $-\dfrac{7}{3}$ *located on a number line?*

(A) -2 *and* -1 (B) -3 *and* -2

(C) -4 *and* -3 (D) -1 *and* 0

10. *Point* P *is at* $(-2, y)$ *and point* Q *is at* $(-2, 5)$. *The distance between* P *and* Q *is 8 units. Find the two possible values of* y.

11. *Simplify:* $9x + 2 + 3x - 7$

Find more at
ViewMath.com/IN-Grade6

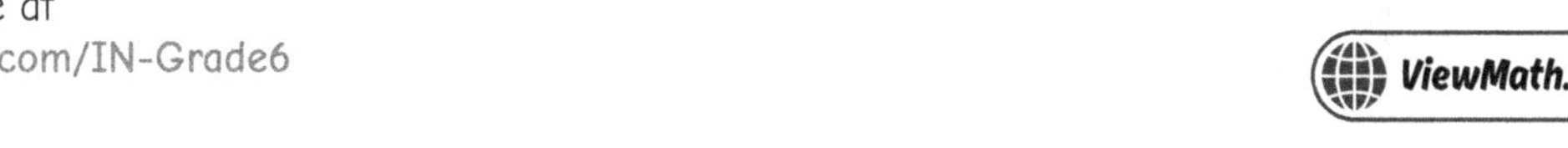

12. Solve: $\dfrac{m}{4} = 7$

(A) $m = 3$

(B) $m = 11$

(C) $m = 28$

(D) $m = 47$

13. Write a real-world situation that can be described by $x \leq 15$.

Your Answer:

14. Which of these numbers is a solution to $x \geq -3$?

(A) -4

(B) -3.5

(C) -3

(D) -100

15. Which table matches the equation $y = x + 4$?

(A) $x : 1, 2, 3 \quad y : 5, 6, 7$

(B) $x : 1, 2, 3 \quad y : 4, 8, 12$

(C) $x : 1, 2, 3 \quad y : 5, 8, 11$

(D) $x : 1, 2, 3 \quad y : 6, 7, 8$

16. A triangle has an area of $45 \ m^2$ and a base of $10 \ m$. What is the height?

Your Answer:

17. A trapezoid has bases 9 in and 15 in and height 6 in. What is the area?

(A) $90 \ in^2$

(B) $72 \ in^2$

(C) $45 \ in^2$

(D) $135 \ in^2$

Find more at
ViewMath.com/IN-Grade6

18. *A box is 8 in long, 6 in wide, and 3 in tall. What is its volume?*

(A) $17\ in^3$

(B) $144\ in^3$

(C) $48\ in^3$

(D) $180\ in^3$

19. *A triangle has vertices $A(1,3)$, $B(5,3)$, and $C(5,7)$. What are the vertices after reflecting across the x-axis?*

(A) $A'(1,-3)$, $B'(5,-3)$, $C'(5,-7)$

(B) $A'(-1,3)$, $B'(-5,3)$, $C'(-5,7)$

(C) $A'(-1,-3)$, $B'(-5,-3)$, $C'(-5,-7)$

(D) $A'(1,3)$, $B'(5,3)$, $C'(5,7)$

20. *A rectangle on the coordinate plane has an area of 56 square units. Its length is 8 units. What is the width?*

(A) 6 units

(B) 7 units

(C) 8 units

(D) 48 units

21. *A triangle has vertices $A(1,2)$, $B(4,2)$, and $C(4,6)$. Reflect the triangle across the x-axis. List all three reflected vertices.*

Your Answer:

22. *What is the area of a circle with radius 5 inches? Use $\pi \approx 3.14$.*

(A) $15.7\ in^2$

(B) $31.4\ in^2$

(C) $78.5\ in^2$

(D) $157\ in^2$

23. Which pair of questions are **both** statistical?

(A) "How many days in a week?" and "How many days in a year?"

(B) "How tall are the trees in the park?" and "How fast can each student run a mile?"

(C) "What is 7×6?" and "How many sides does a triangle have?"

(D) "How many inches in a foot?" and "What is $100 \div 4$?"

24. The data set $12, 14, 15, 15, 16, 17, 18$ has values that range from 12 to 18. What does this tell you about the data?

(A) The center is 15.

(B) The data is very spread out.

(C) The spread (range) is 6.

(D) The data has an outlier.

25. Data: $10, 10, 10, 10, 10$. What is the MAD?

(A) 0

(B) 2

(C) 5

(D) 10

26. A data set has two values that each appear 4 times, and no other value appears more than 3 times. How many modes does this data set have?

(A) 0

(B) 1

(C) 2

(D) 4

27. A box plot shows: min $= 25$, $Q1 = 35$, median $= 50$, $Q3 = 65$, max $= 80$. Find the range and IQR.

Your Answer:

28. *Store A daily sales: median = $500, range = $200. Store B daily sales: median = $500, range = $800. What can you conclude?*

(A) Both stores are equally consistent.

(B) Store A is more consistent in its daily sales.

(C) Store B is more consistent in its daily sales.

(D) Store B has lower typical sales.

29. *The spinner below has 8 equal sections. Find each probability as a fraction in simplest form.*
(a) P(red)   (b) P(blue)   (c) P(not green)

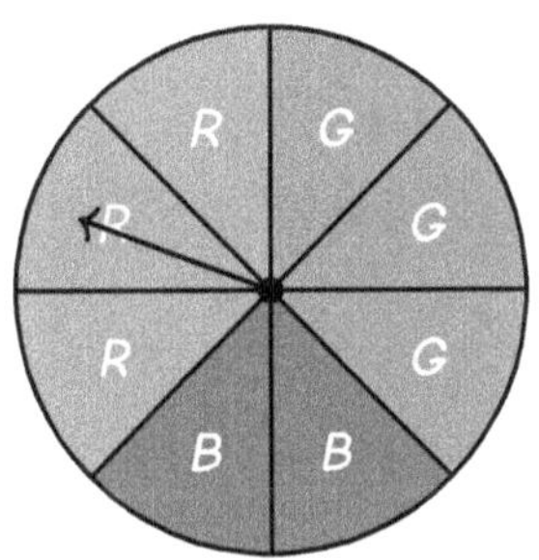

Your Answer:

30. *A bar graph is missing its title. The horizontal axis is labeled "Favorite Sport" and the vertical axis is labeled "Number of Students." Which title best fits this graph?*

(A) Students' Heights in Sixth Grade

(B) Favorite Sports of Sixth Graders

(C) Monthly Rainfall in Our City

(D) Temperature Over One Week

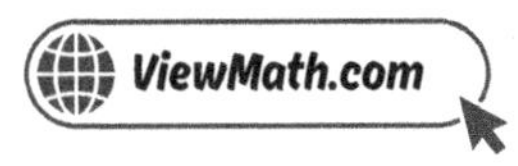

 # End of Practice Test 5 

Great job finishing the test!

 **My Score**

I got _____________ out of 30 questions right.

Check your answers in the **Answer Key** at the back of the book.

Review any questions you missed. That's how we learn!

Check Your Score Online!

Visit **ViewMath Academy** to enter your answers and see which topics you need to review. You can also explore lessons, take quizzes, track your scores, and save your progress!

viewmath.com/score/6.1.IN.20

Or go to *viewmath.com/score* and enter code: 6.1.IN.20

Practice Test 6

✔ 30 Questions

✏ Before You Start ✏

✔ **Read each question carefully** before choosing your answer.

✔ **Show your work** on scratch paper when you need to.

✔ **Skip hard questions** and come back to them later.

✔ **Check your answers** when you're done.

✔ **Take your time** — there's no rush!

★ You've Got This! ★

Do your best and show what you know!

1. A smoothie recipe uses 3 bananas **for every** 4 cups of yogurt. Write the ratio of bananas to yogurt. Then write the ratio of yogurt to bananas.

Your Answer:

2. Brand X sells 5 notebooks for $8.75. Brand Y sells 3 notebooks for $4.50. Which is the better deal?

(A) Brand X at $1.75 each

(B) Brand Y at $1.50 each

(C) Brand X at $1.50 each

(D) Brand Y at $1.75 each

3. Which table matches a graph that passes through $(0,0)$, $(2,3)$, and $(4,6)$?

(A)

x	y
2	3
6	9

(B)

x	y
2	3
6	8

(C)

x	y
3	2
6	9

(D)

x	y
2	4
6	12

4. What is 0.35 written as a percent?

(A) 0.35%

(B) 3.5%

(C) 35%

(D) 350%

5. A race is 10 kilometers. How many centimeters is that?

(A) 100,000

(B) 10,000

(C) 1,000,000

(D) 1,000

6. Mia wants to estimate $4{,}218 \div 7$. She rounds $4{,}218$ to $4{,}200$. What is Mia's estimate?

   (A) 60

   (B) 600

   (C) 602

   (D) 6,000

7. A rope is 12.6 meters long. You cut it into pieces that are each 0.9 meters. How many pieces do you get?

   (A) 1.4

   (B) 13

   (C) 14

   (D) 140

8. What is the LCM of 3 and 9?

   (A) 9

   (B) 3

   (C) 27

   (D) 18

9. Where is $-\dfrac{3}{4}$ located on a number line?

   (A) Between 0 and 1

   (B) Between -2 and -1

   (C) Between -1 and 0

   (D) Between 1 and 2

10. A bird sits at $(1, -2)$ on a coordinate grid. It flies straight up to $(1, 5)$. How many units did it fly?

    (A) 3

    (B) -7

    (C) 7

    (D) 12

11. Which expression is equivalent to $7(k - 3) + 5$?

    (A) $7k - 16$

    (B) $7k + 2$

    (C) $7k - 26$

    (D) $7k - 21$

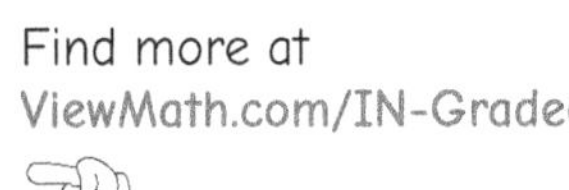
Find more at
ViewMath.com/IN-Grade6

12. The balance scale below is balanced. Each triangle represents the same unknown weight x. Each small square weighs 1 pound. What is the value of x?

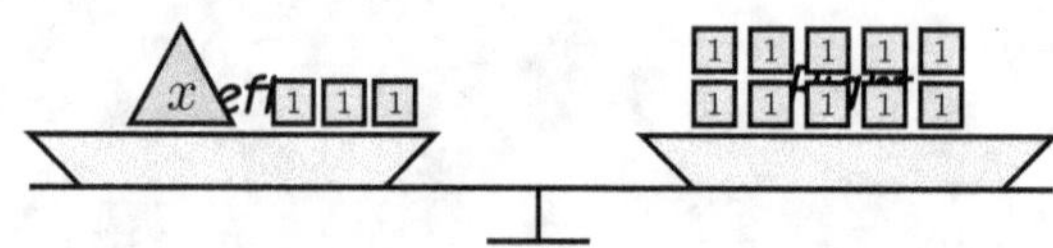

(A) $x = 3$

(B) $x = 7$

(C) $x = 10$

(D) $x = 13$

13. A sign says "No more than 4 guests per room." Write an inequality for the number of guests g.

Your Answer:

14. The pool opens when the temperature is more than 75°F. Which describes the graph of this inequality?

(A) Closed circle at 75, shade right

(B) Open circle at 75, shade right

(C) Closed circle at 75, shade left

(D) Open circle at 75, shade left

15. Which equation represents this relationship: "The number of legs L is 4 times the number of dogs d"?

(A) $d = 4L$

(B) $L = d + 4$

(C) $L = 4d$

(D) $L = d \div 4$

16. A triangle has base 2.4 m and height 5 m. What is the area?

(A) $12\ m^2$

(B) $7.4\ m^2$

(C) $6\ m^2$

(D) $24\ m^2$

Find more at
ViewMath.com/IN-Grade6

ViewMath.com

17. *A trapezoid has bases of 7 m and 11 m and a height of 5 m. What is the area?*

Your Answer

18. *A swimming pool is 25 m long, 10 m wide, and 2 m deep. How many cubic meters of water does it hold when full?*

Your Answer

19. *A rectangle has vertices $(-3, 4)$, $(5, 4)$, $(5, -2)$, and $(-3, -2)$. What is the length of the horizontal side?*

(A) 5 units

(B) 6 units

(C) 8 units

(D) 3 units

20. *A rectangle with vertices $(1, 2)$, $(7, 2)$, $(7, 6)$, and $(1, 6)$ has a right triangle cut from it with vertices $(1, 2)$, $(7, 2)$, and $(7, 6)$. What is the area of the remaining piece?*

Your Answer

21. *Point $M(-2, 7)$ moves to $M'(4, 7)$. What transformation occurred?*

(A) Reflection across the x-axis

(B) Reflection across the y-axis

(C) Translation 6 units right

(D) Translation 6 units left

22. *A circle has a diameter of 18 inches. What is its area? Use $\pi \approx 3.14$.*

Your Answer

Find more at
ViewMath.com/IN-Grade6

23. Lena changes a non-statistical question into a statistical one. The original was "How many ounces are in a cup?" Which revision makes it statistical?

(A) How many ounces are in 3 cups?

(B) How many cups of water does each student drink per day?

(C) How many ounces are in a gallon?

(D) What is 8×2?

24. Test scores: $68, 70, 72, 73, 74, 75, 76, 98$. Where is the center of this data?

(A) Around 68–70

(B) Around 73–75

(C) Around 83

(D) Around 98

25. Data: $20, 22, 24, 26, 28$. The mean is 24. What is the MAD?

(A) 2

(B) 2.4

(C) 4

(D) 8

26. Which display is best for showing favorite colors of students?

(A) Histogram

(B) Dot plot

(C) Frequency table or bar graph

(D) Box plot

27. Which data set has the five-number summary: $5, 10, 15, 20, 25$?

(A) $5, 10, 15, 20, 25$

(B) $5, 8, 10, 15, 18, 20, 22, 25$

(C) $5, 7, 10, 12, 15, 18, 20, 23, 25$

(D) Any of these could have that five-number summary.

Find more at
ViewMath.com/IN-Grade6

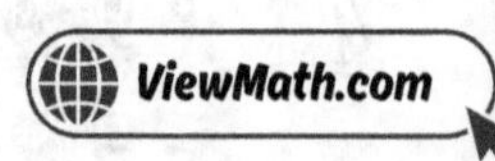

28. *Daily steps for Week 1: median = 8,000, IQR = 2,000. Week 2: median = 9,500, IQR = 1,200. Which week shows a higher typical step count and more consistent activity?*

(A) *Week 1 for both*

(B) *Week 2 for both*

(C) *Week 1 higher, Week 2 more consistent*

(D) *Week 2 higher, Week 1 more consistent*

29. *A fair coin is flipped once. What is the probability of landing on heads?*

(A) $\dfrac{1}{4}$

(B) $\dfrac{1}{3}$

(C) $\dfrac{1}{2}$

(D) 1

30. *A frequency table shows that 5 students chose pizza, 8 chose tacos, 3 chose pasta, and 4 chose salad. How many students were surveyed in all?*

(A) 16

(B) 18

(C) 20

(D) 24

Find more at
ViewMath.com/IN-Grade6

 # End of Practice Test 6 

Great job finishing the test!

 **My Score**

I got _____________ out of 30 questions right.

*Check your answers in the **Answer Key** at the back of the book.*

Review any questions you missed. That's how we learn!

Check Your Score Online!

Visit **ViewMath Academy** to enter your answers and see which topics
you need to review. You can also explore lessons, take quizzes, track
your scores, and save your progress!

viewmath.com/score/6.1.IN.21

Or go to viewmath.com/score and enter code: 6.1.IN.21

7

Practice Test 7

☑ *30 Questions*

✏️ Before You Start ✏️

- ✓ **Read each question carefully** before choosing your answer.
- ✓ **Show your work** on scratch paper when you need to.
- ✓ **Skip hard questions** and come back to them later.
- ✓ **Check your answers** when you're done.
- ✓ **Take your time** — there's no rush!

★ You've Got This! ★

Do your best and show what you know!

1. *A lemonade stand sells 7 cups of lemonade **for each** 2 cups of iced tea. If they sold 21 cups of lemonade, how many cups of iced tea did they sell?*

Your Answer:

2. *A garden hose fills a 150-gallon tank in 5 hours. What is the unit rate?*

(A) 25 gallons per hour

(B) 30 gallons per hour

(C) 35 gallons per hour

(D) 750 gallons per hour

3. *A ratio graph passes through $(4, 10)$ and the origin. List two other points on this line.*

Your Answer:

4. *A class has 30 students. 70% passed a quiz. How many students passed?*

(A) 7

(B) 21

(C) 23

(D) 70

5. *A recipe calls for 3 pints of broth. You have a 1-quart container. How many quarts is 3 pints? (1 quart $= 2$ pints) Will one container be enough?*

Your Answer:

6. *What is $936 \div 4$?*

(A) 231

(B) 234

(C) 239

(D) 2,304

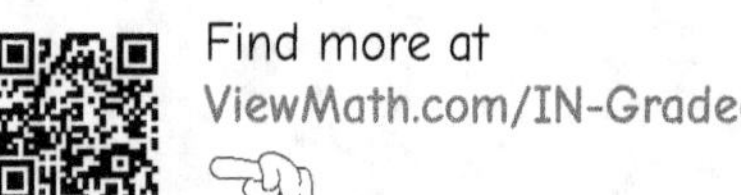
Find more at
ViewMath.com/IN-Grade6

7. Compute $0.288 \div 0.12$.

8. What is the GCF of 60 and 84?

(A) 4

(B) 6

(C) 12

(D) 420

9. Which of the following is a rational number?

(A) $\sqrt{2}$

(B) π

(C) $-\dfrac{3}{4}$

(D) $\sqrt{5}$

10. Two traffic lights are at $(-3, 5)$ and $(4, 5)$ along the same street on a city grid. How many blocks apart are they?

(A) 1

(B) 5

(C) 7

(D) 9

11. Factor using the distributive property: $15n + 10$

12. Is $x = 5$ a solution to $3x = 18$?

(A) Yes, because $3 + 5 = 18$

(B) Yes, because $3 \times 5 = 18$

(C) No, because $3 \times 5 = 15$

(D) No, because $3 + 5 = 8$

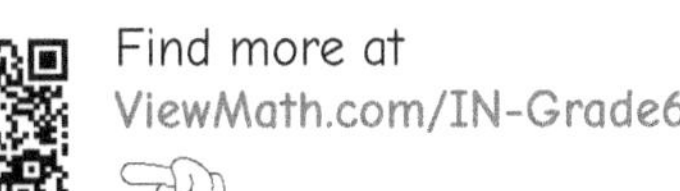
Find more at
ViewMath.com/IN-Grade6

13. A movie is rated PG-13, meaning you must be over 13 to watch without a parent. Which inequality represents ages that can watch without a parent?

(A) $a < 13$

(B) $a > 13$

(C) $a \leq 13$

(D) $a = 13$

14. Which of the following graphs represents $x \leq -1$?

(A) Open circle at -1, shade right

(B) Closed circle at -1, shade right

(C) Open circle at -1, shade left

(D) Closed circle at -1, shade left

15. A movie streaming service costs \$5 per month plus a \$10 sign-up fee. Write an equation for the total cost c after m months. Find the cost after 6 months.

Your Answer:

16. A triangle has base 8 in and height 5 in. What is its area?

(A) $13\ in^2$

(B) $40\ in^2$

(C) $20\ in^2$

(D) $26\ in^2$

17. *What is the area of a parallelogram with base 9 cm and height 5 cm?*

(A) $14\ cm^2$

(B) $45\ cm^2$

(C) $22.5\ cm^2$

(D) $28\ cm^2$

18. *A rectangular prism has length $\frac{3}{4}$ ft, width 2 ft, and height 4 ft. What is the volume?*

(A) $6\ ft^3$

(B) $3\ ft^3$

(C) $24\ ft^3$

(D) $\frac{3}{2}\ ft^3$

19. *Point B is at $(-6, -4)$. When reflected across the y-axis, where does B' land?*

(A) $(-6, 4)$

(B) $(6, 4)$

(C) $(6, -4)$

(D) $(4, -6)$

20. *An L-shaped figure has vertices $(0,0)$, $(10,0)$, $(10,4)$, $(6,4)$, $(6,8)$, and $(0,8)$. What is the total area?*

Your Answer:

21. A triangle has vertices $A(2, 1)$, $B(6, 1)$, and $C(4, 5)$. Each vertex is translated 3 units left. What are the new coordinates of B'?

(A) $(3, 1)$

(B) $(9, 1)$

(C) $(6, -2)$

(D) $(6, 4)$

22. A bike wheel has a radius of 13 inches. How far does the bike travel in one full rotation of the wheel? Use $\pi \approx 3.14$

(A) 40.82 in

(B) 81.64 in

(C) 163.28 in

(D) 530.66 in

23. Is the following question statistical or not statistical? Explain.

"How many hours of sleep did each student in Ms. Rivera's class get last night?"

Your Answer:

24. Look at the dot plot below.

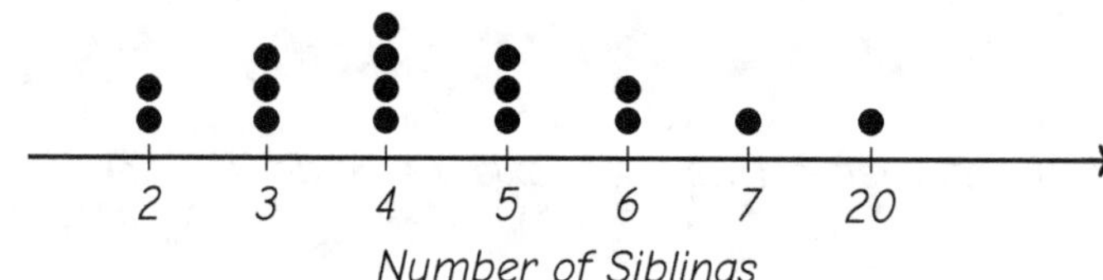

Which value is an outlier?

(A) 2

(B) 4

(C) 7

(D) 20

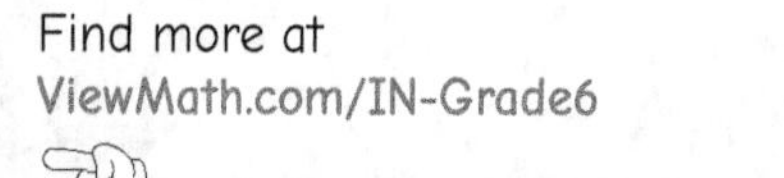
Find more at
ViewMath.com/IN-Grade6

ViewMath.com

25. *Data:* 100, 105, 110, 115, 120, 125, 130. *Find the IQR.*

Your Answer:

26. *A data set is:* 5, 5, 5, 5, 5. *What is the mode?*

(A) 0

(B) 1

(C) 5

(D) *There is no mode.*

27. *The two box plots below represent daily sales (in dollars) at two stores over the same month.*

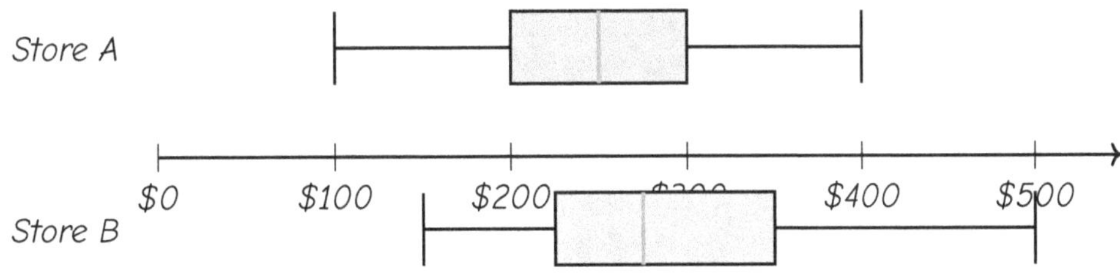

Compare the two stores. Which store has higher typical sales? Which has more predictable sales? Use the box plots to support your answers.

Your Answer:

28. *Which two measures together best describe the "center" and "spread" of a data set?*

(A) *Mean and mode*

(B) *Median and IQR*

(C) *Range and maximum*

(D) *Minimum and maximum*

29. Event A has a probability of $\frac{2}{5}$ and Event B has a probability of $\frac{3}{4}$. Which statement is true?

   (A) Event A is more likely than Event B   (B) Event B is more likely than Event A

   (C) Both events are equally likely   (D) Neither event can happen

30. A double bar graph compares boys and girls who participated in a fun run. The bars show 24 boys and 30 girls. What fraction of all participants were boys?

   (A) $\frac{24}{30}$ (B) $\frac{4}{9}$

   (C) $\frac{5}{9}$ (D) $\frac{1}{2}$

 *End of Practice Test 7* 

Great job finishing the test!

 **My Score**

I got _____________ out of 30 questions right.

Check your answers in the Answer Key at the back of the book.

Review any questions you missed. That's how we learn!

Check Your Score Online!

*Visit **ViewMath Academy** to enter your answers and see which topics you need to review. You can also explore lessons, take quizzes, track your scores, and save your progress!*

viewmath.com/score/6.1.IN.22

Or go to viewmath.com/score and enter code: 6.1.IN.22

Practice Test 8

30 Questions

✏ Before You Start ✏

- ✔ **Read each question carefully** before choosing your answer.
- ✔ **Show your work** on scratch paper when you need to.
- ✔ **Skip hard questions** and come back to them later.
- ✔ **Check your answers** when you're done.
- ✔ **Take your time** — there's no rush!

★ You've Got This! ★

Do your best and show what you know!

1. The ratio of cats to dogs at a pet store is 3 : 4. Each part in the tape diagram represents 2 animals. How many dogs are there?

(A) 6

(B) 8

(C) 3

(D) 4

2. The graph shows the number of laps a swimmer completes over time.

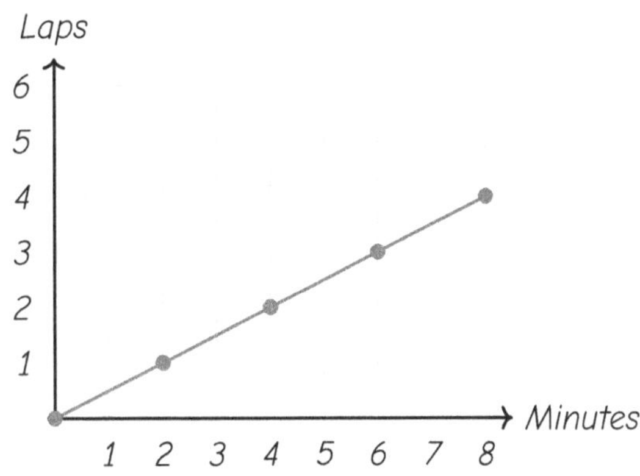

Part A: What is the swimmer's unit rate in laps per minute?

Part B: How many laps will the swimmer complete in 14 minutes?

Your Answer:

3. True or false: a graph of equivalent ratios can be a curved line. Explain.

Your Answer:

4. Which of the following is 100% as a fraction?

(A) $\dfrac{1}{100}$

(B) $\dfrac{100}{10}$

(C) $\dfrac{100}{100}$

(D) $\dfrac{10}{100}$

5. A truck can carry 3 tons. A shipment weighs 5,500 pounds. Can the truck carry the shipment? (1 ton = 2,000 pounds)

Your Answer:

6. The table shows how many pieces of fruit a warehouse packs into crates.

Fruit	Total Pieces	Pieces per Crate
Apples	2,688	16
Oranges	1,575	25
Bananas	1,890	15

Which fruit requires the **most** crates?

(A) Apples, because $2,688 \div 16 = 168$ crates

(B) Bananas, because $1,890 \div 15 = 126$ crates

(C) Oranges, because $1,575 \div 25 = 63$ crates

(D) Bananas, because 1,890 is the largest two-digit divisor

7. A student computed 2.4×0.3 and got 7.2. What mistake did the student most likely make?

(A) The student added instead of multiplying

(B) The student counted only one decimal place instead of two

(C) The student forgot to place any decimal point

(D) The student divided instead of multiplying

8. Which of the following is **not** a common factor of 24 and 36?

(A) 4

(B) 8

(C) 6

(D) 12

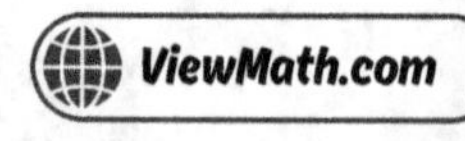

9. *Which number is farthest from 0 on a number line?*

(A) 0.8

(B) $-\dfrac{3}{2}$

(C) 1.1

(D) $-\dfrac{4}{5}$

10. **Part A:** *Find the horizontal distance between* $(-5, 3)$ *and* $(2, 3)$.

Part B: *Find the vertical distance between* $(2, 3)$ *and* $(2, -4)$.

Part C: *If you walk from* $(-5, 3)$ *to* $(2, 3)$ *and then from* $(2, 3)$ *to* $(2, -4)$, *what is the total distance you walk?*

11. *The table below shows values for two expressions. Based on the table, is Expression A equivalent to Expression B?*

x	Expression A: $2(x+3)$	Expression B: $2x+6$
0	6	6
1	8	8
5	16	16
10	26	26

(A) *No — they only agree for small values of* x

(B) *Yes — the distributive property shows* $2(x+3) = 2x+6$ *for all* x

(C) *No — the table shows different values*

(D) *Cannot be determined from the table*

12. *Solve:* $\dfrac{t}{6} = 5$

(A) $t = 11$

(B) $t = 1$

(C) $t = 30$

(D) $t = 56$

13. Name three solutions to the inequality $m \geq 6$.

Your Answer:

14. Two number lines are shown below. Write the inequality for each.

Graph A:

Graph B:

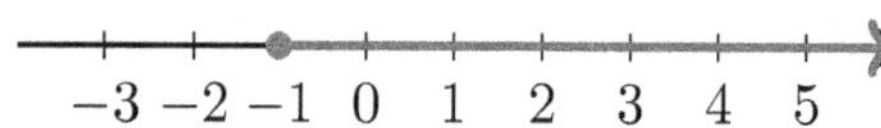

Your Answer:

15. A bathtub drains at 3 gallons per minute. It starts with 45 gallons. Write an equation for the water w remaining after t minutes. When is the tub empty?

Your Answer:

16. A triangle has base $\frac{3}{4}$ ft and height $\frac{2}{3}$ ft. What is its area?

Your Answer:

17. A parallelogram has an area of 96 ft^2 and a height of 8 ft. What is the base?

Your Answer:

18. A toy box is 3 ft long, 2 ft wide, and 2 ft tall. What is the volume?

(A) $7 \, ft^3$

(B) $12 \, ft^3$

(C) $6 \, ft^3$

(D) $24 \, ft^3$

19. A triangle has vertices $P(2, 4)$, $Q(6, 4)$, and $R(6, 1)$. Reflect the triangle across the y-axis. What are the coordinates of P'?

20. The L-shaped figure below is drawn on a grid. What is its area?

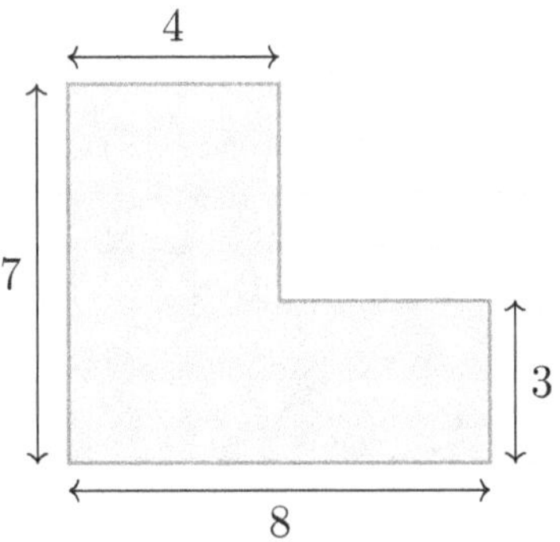

(A) 40 square units

(B) 32 square units

(C) 56 square units

(D) 24 square units

21. Point $G(-3, 5)$ is reflected across the y-axis and then translated 2 units down. What are the final coordinates?

22. A circle has a diameter of 10 cm. What is its circumference? Use $\pi \approx 3.14$

(A) 15.7 cm

(B) 31.4 cm

(C) 62.8 cm

(D) 314 cm

23. What makes a question statistical?

(A) It can be answered with a number.

(B) It is about a math topic.

(C) The answers are expected to vary.

(D) It has exactly one correct answer.

24. Data: $10, 12, 14, 15, 16, 18, 50$. Identify any outliers and explain their effect on the center.

Your Answer:

25. Data: $50, 55, 60, 65, 70$. The mean is 60. Find the MAD.

Your Answer:

26. You survey 15 classmates about their favorite fruit. Which display should you start with?

(A) Histogram, because it groups data into intervals.

(B) Frequency table, because you need to count each category first.

(C) Box plot, because you need the five-number summary.

(D) Dot plot, because it shows individual numerical values.

27. A box plot has: min = 60, Q1 = 70, median = 75, Q3 = 85, max = 100. What percent of data falls between 70 and 85?

Your Answer:

Find more at
ViewMath.com/IN-Grade6

28. Which pair of measures should you report when outliers are present?

(A) Mean and range

(B) Mode and maximum

(C) Median and IQR

(D) Mean and IQR

29. A bag contains 4 green marbles, 6 red marbles, and 10 blue marbles. What is the probability of randomly drawing a green marble? Write your answer as a percent.

Your Answer

30. In a frequency table, the colors chosen by students are: red (12), blue (7), green (7), yellow (4). How many more students chose red than yellow?

(A) 4

(B) 5

(C) 7

(D) 8

 # End of Practice Test 8 

Great job finishing the test!

 My Score

I got _____________ out of 30 questions right.

Check your answers in the **Answer Key** at the back of the book.

💡 Review any questions you missed. That's how we learn!

📊 Check Your Score Online!

Visit **ViewMath Academy** to enter your answers and see which topics you need to review. You can also explore lessons, take quizzes, track your scores, and save your progress!

viewmath.com/score/6.1.IN.23

Or go to viewmath.com/score and enter code: 6.1.IN.23

Practice Test 9

☑ 30 Questions

✏ Before You Start ✏

✔ **Read each question carefully** before choosing your answer.

✔ **Show your work** on scratch paper when you need to.

✔ **Skip hard questions** and come back to them later.

✔ **Check your answers** when you're done.

✔ **Take your time** — there's no rush!

★ You've Got This! ★

Do your best and show what you know!

1. The ratio of red to blue to green beads is $1:3:2$. There are 18 beads total. How many blue beads are there?

   (A) 3

   (B) 6

   (C) 9

   (D) 12

2. A printer prints 120 pages in 4 minutes. What is the unit rate?

   (A) 4 pages per minute

   (B) 30 pages per minute

   (C) 120 pages per minute

   (D) 480 pages per minute

3. Two ratio graphs both pass through the origin. Line A goes through $(1, 4)$ and Line B goes through $(1, 3)$. Which line is steeper?

   (A) Line A

   (B) Line B

   (C) They have the same steepness.

   (D) Cannot be determined.

4. A test has 100 questions. Emma got 88 correct. What percent did she get right?

   (A) 12%

   (B) 8.8%

   (C) 88%

   (D) 78%

5. Convert 3 gallons to pints. (1 gallon $= 8$ pints)

   (A) 16

   (B) 24

   (C) 32

   (D) 11

Find more at
ViewMath.com/IN-Grade6

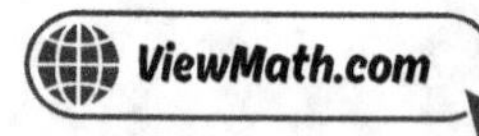

6. A school orders 1,260 pencils. Each box holds 18 pencils. How many boxes is that?

(A) 7

(B) 700

(C) 70

(D) 17

7. Emma walks 1.25 miles to school and the same distance home. She does this 5 days a week. How many miles does she walk in one week?

(A) 6.25

(B) 12.5

(C) 10.25

(D) 13.5

8. Marcus listed the factors of 48 and 72 in the table below, but left some blanks.

Factors of 48	1, 2, 3, 4, 6, 8, 12, 16, 24, 48
Factors of 72	1, 2, 3, 4, 6, 8, 9, 12, 18, 24, 36, 72

Part A: List all the common factors of 48 and 72.

Part B: What is the GCF of 48 and 72?

Part C: A baker made 48 cookies and 72 brownies. She wants to pack them into identical boxes with no treats left over. Using your answer from Part B, how many boxes can she fill, and what goes in each box?

Your Answer

9. Which number is closest to 0 on a number line?

(A) $-\dfrac{2}{3}$

(B) $\dfrac{3}{4}$

(C) $-\dfrac{1}{5}$

(D) $\dfrac{5}{6}$

Find more at
ViewMath.com/IN-Grade6

10. *Three points are plotted on the coordinate plane below.*

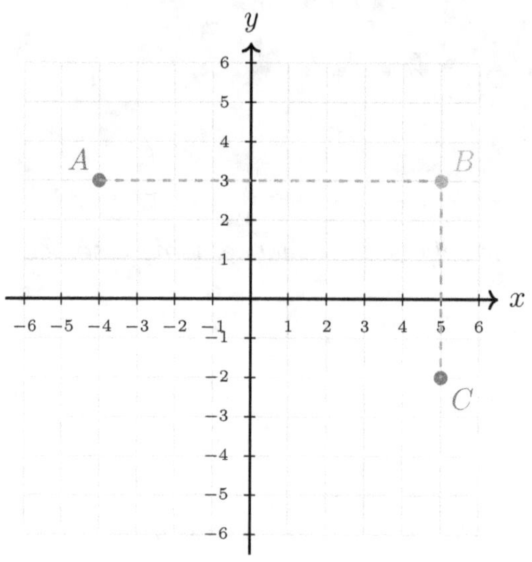

Part A: *What is the distance from A to B?*

Part B: *What is the distance from B to C?*

Part C: *If you walk from A to B and then from B to C, what is the total distance?*

Your Answer

---

11. *Which expression is equivalent to $9w - 4w + 6 - 6$?*

(A) $5w$

(B) $5w + 12$

(C) $13w$

(D) 5

---

12. *Solve:* $x + 9 = 15$

(A) $x = 24$

(B) $x = 6$

(C) $x = 4$

(D) $x = 9$

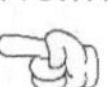

Find more at
ViewMath.com/IN-Grade6

13. Which inequality represents "you need more than \$25 to buy the video game"?

(A) $d \leq 25$

(B) $d < 25$

(C) $d > 25$

(D) $d \geq 25$

14. When graphing $x \leq 6$ on a number line, which direction do you shade?

(A) To the right

(B) To the left

(C) Both directions

(D) No shading needed

15. On a graph of $y = 4x$, which ordered pair is on the line?

(A) $(2, 6)$

(B) $(3, 12)$

(C) $(4, 12)$

(D) $(5, 25)$

16. A triangle and a rectangle both have a base of 6 cm and a height of 10 cm. How do their areas compare?

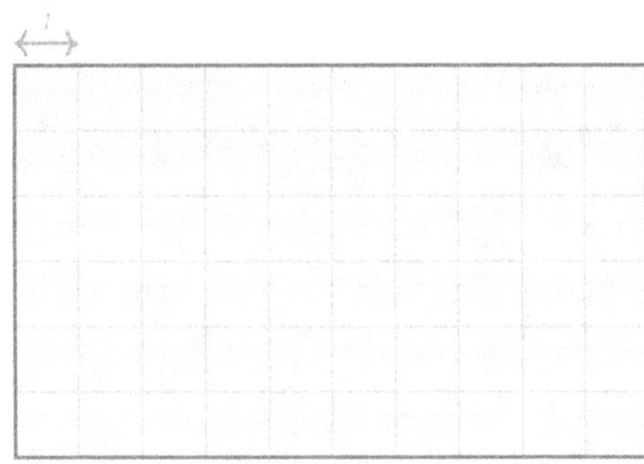

(A) The triangle has twice the area of the rectangle.

(B) They have the same area.

(C) The triangle has half the area of the rectangle.

(D) The rectangle has three times the area of the triangle.

17. *A parallelogram and a rectangle both have a base of 7 m and a height of 4 m. Which statement is true?*

(A) *The rectangle has a greater area.*

(B) *The parallelogram has a greater area.*

(C) *They have the same area.*

(D) *You cannot compare them.*

18. *A rectangular prism has a volume of 120 cm³. Its length is 10 cm and width is 4 cm. What is the height?*

(A) 3 cm

(B) 12 cm

(C) 6 cm

(D) 30 cm

19. *After reflecting $(2, 5)$ across the x-axis, how far is the reflected point from the original?*

(A) 2 units

(B) 5 units

(C) 7 units

(D) 10 units

20. *A composite figure is made of a 5×4 rectangle and a right triangle with base 5 and height 3, placed on top. What is the total area?*

(A) 27.5 square units

(B) 35 square units

(C) 20 square units

(D) 32.5 square units

Find more at
ViewMath.com/IN-Grade6

ViewMath.com

21. After a translation, which of the following is always true?

(A) The shape gets larger.

(B) The shape flips to a mirror image.

(C) The shape changes its angles.

(D) The shape keeps the same size, shape, and orientation.

22. What does the number π (pi) represent?

(A) The radius divided by the diameter

(B) The ratio of a circle's circumference to its diameter

(C) The area of any circle with radius 1

(D) The diameter divided by the circumference

23. Which question would help you collect data to make a dot plot?

(A) What is the capital of Texas?

(B) How many books did each student read this month?

(C) What is $5 + 5$?

(D) How many sides does a hexagon have?

24. Which term describes a data value that is much higher or lower than the rest of the data?

(A) Cluster

(B) Gap

(C) Outlier

(D) Peak

Find more at
ViewMath.com/IN-Grade6

25. Look at the two dot plots below.

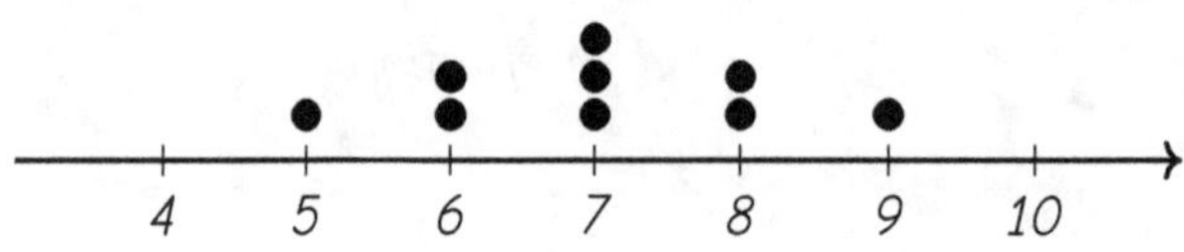

Data Set A

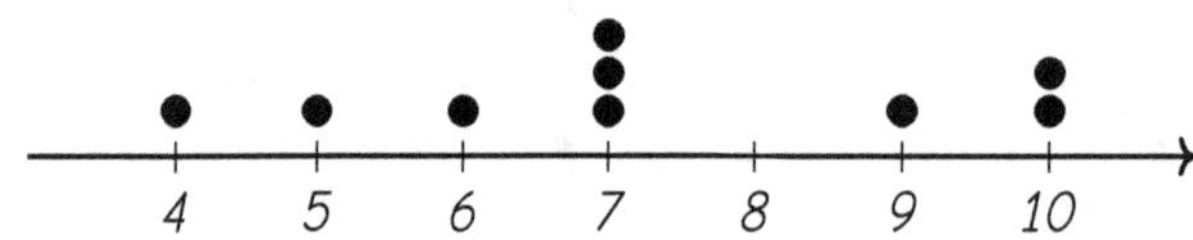

Data Set B

Which data set has a larger range?

(A) Data Set A

(B) Data Set B

(C) They have the same range.

(D) Cannot be determined.

26. You want to know whether to display data as a dot plot, histogram, or frequency table. Which question should you ask first?

(A) What is the mean of the data?

(B) How many data points are there, and is the data numerical or categorical?

(C) What color should the bars be?

(D) Is the data symmetric or skewed?

Find more at
ViewMath.com/IN-Grade6

27. Look at the two box plots below showing test scores for two classes.

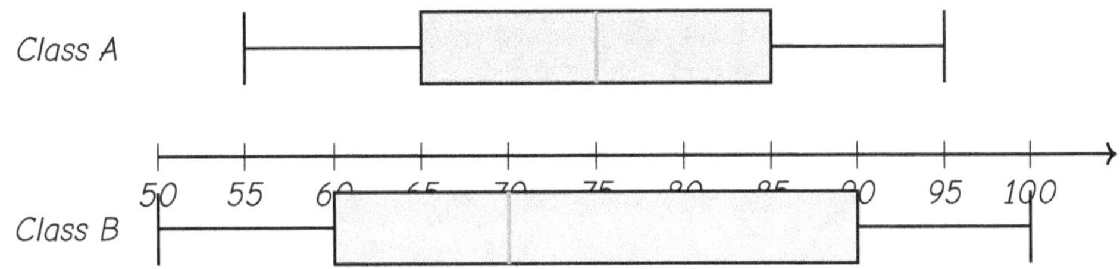

Which class has more consistent scores?

(A) Class A, because its box is narrower.

(B) Class B, because its box is wider.

(C) Both are equally consistent.

(D) Class B, because its median is higher.

28. Which measure of spread is affected by every single data value?

(A) IQR

(B) Range

(C) Median

(D) Mode

29. A jar holds 10 jellybeans: 4 cherry, 4 grape, and 2 lemon. Which event is **equally likely** as drawing a cherry jellybean?

(A) Drawing a lemon jellybean

(B) Drawing a grape jellybean

(C) Not drawing a jellybean

(D) Drawing a cherry or lemon jellybean

30. A double bar graph shows the number of apples and oranges sold each day. On Monday, 15 apples and 20 oranges were sold. On Tuesday, 25 apples and 10 oranges were sold. On which day were more total fruits sold, and how many more?

Your Answer:

Find more at
ViewMath.com/IN-Grade6

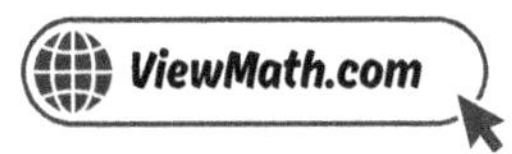

 **End of Practice Test 9** 

Great job finishing the test!

 **My Score**

I got _____________ out of 30 questions right.

Check your answers in the **Answer Key** at the back of the book.

💡 Review any questions you missed. That's how we learn!

📊 **Check Your Score Online!**

Visit **ViewMath Academy** to enter your answers and see which topics you need to review. You can also explore lessons, take quizzes, track your scores, and save your progress!

viewmath.com/score/6.1.IN.24

Or go to viewmath.com/score and enter code: 6.1.IN.24

Practice Test 10

 30 Questions

✏ Before You Start ✏

- ✔ **Read each question carefully** before choosing your answer.
- ✔ **Show your work** on scratch paper when you need to.
- ✔ **Skip hard questions** and come back to them later.
- ✔ **Check your answers** when you're done.
- ✔ **Take your time** — there's no rush!

 ⭐ You've Got This!  ⭐

Do your best and show what you know!

1. The tape diagram below shows the ratio of apple juice to orange juice in a drink.

Apple:

Orange:

If each part equals 4 ounces, how many total ounces of drink are there?

(A) 12

(B) 20

(C) 32

(D) 8

2. The table below shows the prices at two stores for packs of pencils.

	Number of Pencils	Price
Store A	10	$4.50
Store B	8	$3.20

Which store has the lower unit price per pencil?

(A) Store A at $0.45 per pencil

(B) Store B at $0.40 per pencil

(C) Store A at $0.40 per pencil

(D) They have the same unit price.

3. Juice costs $2 per bottle. Which point would **not** appear on the graph of this ratio?

(A) $(1, 2)$

(B) $(3, 6)$

(C) $(4, 10)$

(D) $(5, 10)$

4. Order from least to greatest: $\frac{1}{3}$, 30%, 0.35.

(A) 30%, $\frac{1}{3}$, 0.35

(B) $\frac{1}{3}$, 30%, 0.35

(C) 0.35, 30%, $\frac{1}{3}$

(D) 30%, 0.35, $\frac{1}{3}$

Find more at
ViewMath.com/IN-Grade6

ViewMath.com

5. Which unit conversion uses a ratio?

(A) $5 + 12 = 17$ inches

(B) $\dfrac{5\ ft}{1} \times \dfrac{12\ in}{1\ ft} = 60\ in$

(C) $5 - 12 = -7$ inches

(D) $\dfrac{12}{5} = 2.4\ feet$

6. The long division below for $1{,}575 \div 5$ is partially completed. A digit in the quotient is missing.

$$
\begin{array}{r}
3\ \square\ 5 \\
5\,\overline{\smash{\big)}\,1\,5\,7\,5} \\
-1\,5 \\
\hline
7 \\
-5 \\
\hline
2\,5 \\
-2\,5 \\
\hline
0
\end{array}
$$

What digit belongs in the box? Explain how you know.

Your Answer:

7. Compute $25.1 - 9.38$.

Your Answer:

8. Two lights blink at regular intervals. One blinks every 4 seconds and the other blinks every 7 seconds. They both blink at time 0. After how many seconds will they blink at the same time again?

Your Answer:

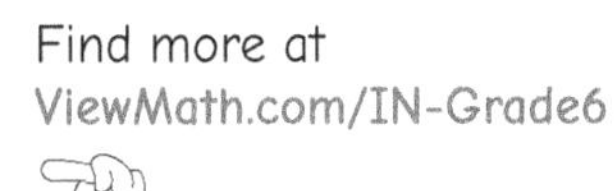
Find more at
ViewMath.com/IN-Grade6

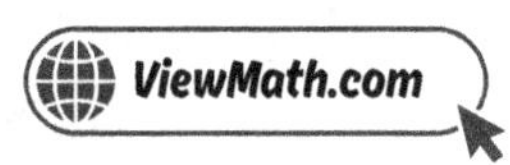

9. Between which two consecutive integers is $-\dfrac{5}{3}$ located on a number line?

   (A) -1 and 0 (B) -2 and -1

   (C) -3 and -2 (D) 0 and 1

10. What is the distance between $(4, -6)$ and $(4, -1)$?

    (A) -7 (B) -5

    (C) 5 (D) 7

11. A student's work simplifying $3(2x + 4) + x$ is shown below. Find and correct the error.

    Step 1: $3(2x + 4) + x$

    Step 2: $6x + 4 + x$

    Step 3: $7x + 4$

    Your Answer:

12. Solve: $12p = 84$

    (A) $p = 72$ (B) $p = 96$

    (C) $p = 7$ (D) $p = 6$

13. Look at the elevator sign below. Write an inequality for the maximum number of people p and the maximum weight w.

ELEVATOR CAPACITY

Maximum: 12 persons

Maximum weight: 2,000 lbs

Your Answer

14. Which value can you use to test that the graph of $x > -5$ is correct?

(A) $x = -5$ (it should make the inequality true)   (B) $x = -6$ (it should make the inequality true)

(C) $x = 0$ (it should make the inequality true)   (D) $x = -10$ (it should make the inequality true)

15. Which table shows a relationship where y does NOT depend on x?

(A) $x : 1, 2, 3$ $y : 3, 6, 9$ (B) $x : 1, 2, 3$ $y : 7, 7, 7$

(C) $x : 1, 2, 3$ $y : 2, 4, 6$ (D) $x : 1, 2, 3$ $y : 5, 10, 15$

16. Find the area of a triangle with base 16 cm and height 11 cm.

Your Answer

17. A parallelogram has base 11 cm and height 3 cm. Maria says the area is 16.5 cm^2. What did Maria do wrong?

(A) She divided by 2 instead of multiplying.   (B) She added the base and height.

(C) She multiplied correctly; 16.5 cm^2 is correct.   (D) She used the wrong unit.

Find more at
ViewMath.com/IN-Grade6

18. An L-shaped solid is formed by joining two rectangular prisms as shown. What is the total volume?

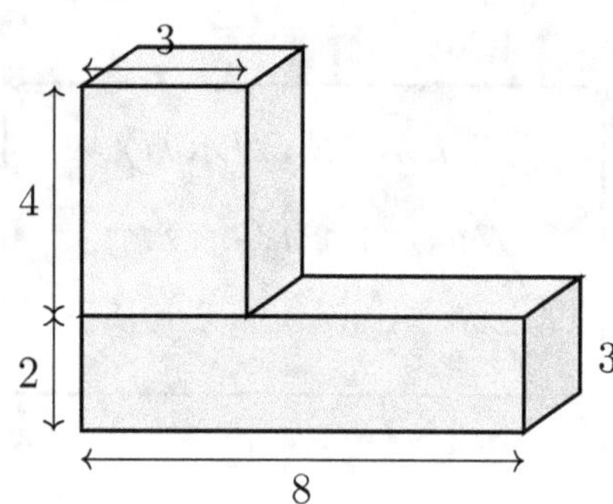

All depths are 3 units.

Your Answer:

19. Which statement about reflections is true?

(A) A reflection changes the size of a shape.

(B) A reflected point is the same distance from the axis as the original.

(C) Reflecting across the x-axis changes the x-coordinate.

(D) Reflecting across the y-axis changes the y-coordinate.

20. To find the area of an irregular polygon on the coordinate plane, you can:

(A) Multiply all the coordinates together.

(B) Break it into rectangles and triangles, then add the areas.

(C) Count only the vertices.

(D) Subtract the perimeter from the largest coordinate.

21. When you reflect a point across the x-axis, which coordinate changes sign?

(A) The x-coordinate

(B) The y-coordinate

(C) Both coordinates

(D) Neither coordinate

Find more at
ViewMath.com/IN-Grade6

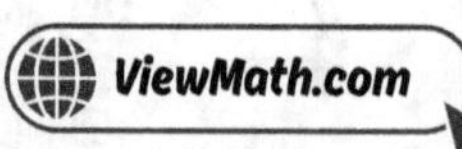

22. A circle has a diameter of 8 feet. What is its circumference? Use $\pi \approx 3.14$.

(A) 12.56 ft

(B) 25.12 ft

(C) 50.24 ft

(D) 200.96 ft

23. Is the following question statistical or not statistical?
"How many text messages does each student send per day?"

24. Look at the two data displays below.

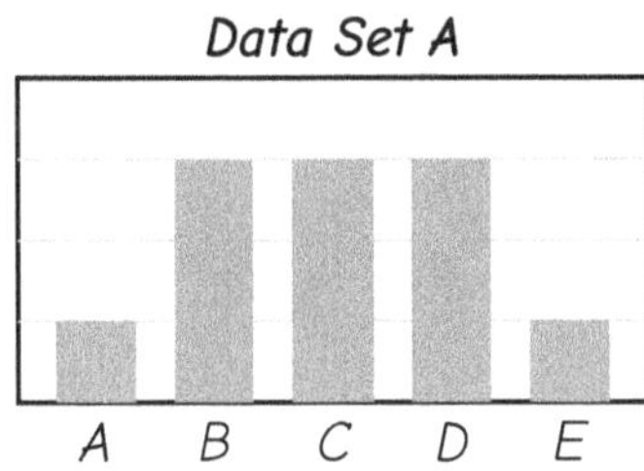

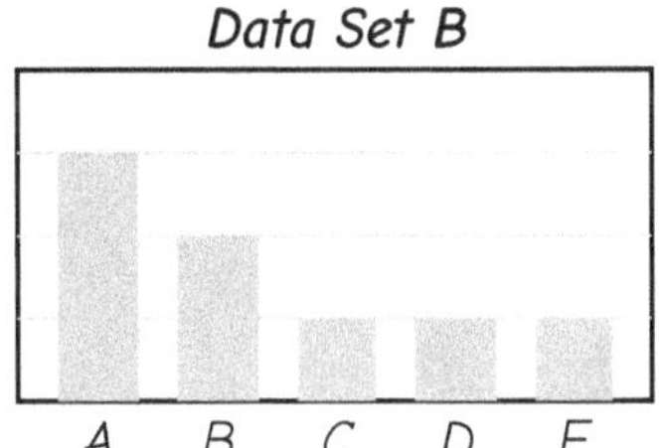

Which data set is skewed?

(A) Data Set A is skewed right and Data Set B is symmetric.

(B) Data Set A is symmetric and Data Set B is skewed right.

(C) Both are symmetric.

(D) Both are skewed.

25. If every value in a data set is increased by 5, what happens to the range?

(A) It increases by 5.

(B) It decreases by 5.

(C) It stays the same.

(D) It doubles.

26. A histogram shows minutes of exercise: 0–14 (3), 15–29 (8), 30–44 (12), 45–59 (7). How many people exercised less than 30 minutes?

Your Answer:

27. In a box plot, the "box" stretches from —

(A) the minimum to the maximum

(B) Q1 to Q3

(C) the median to the maximum

(D) 0 to the median

28. Two box plots are shown on the same number line.

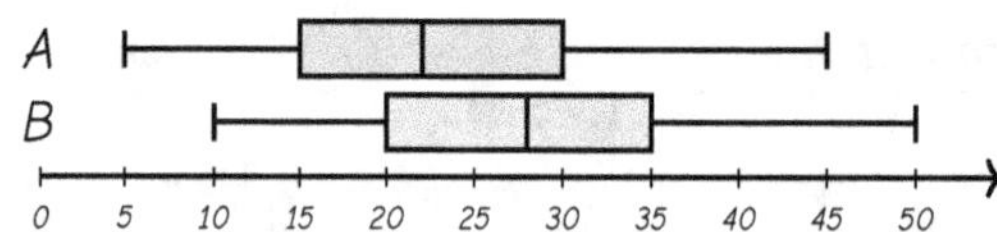

Which data set has a greater IQR?

(A) A (IQR = 15)

(B) B (IQR = 15)

(C) They have the same IQR.

(D) Cannot be determined.

29. A bag has 5 orange balls, 3 purple balls, and 2 yellow balls. What is the probability of **not** drawing a yellow ball?

(A) $\dfrac{2}{10}$

(B) $\dfrac{5}{10}$

(C) $\dfrac{8}{10}$

(D) $\dfrac{3}{10}$

Find more at
ViewMath.com/IN-Grade6

ViewMath.com

30. A frequency table shows test scores. The title of the table is "Math Quiz Scores for Period 3." What does the title tell you?

(A) The scores are from a science test

(B) The scores are from every class in the school

(C) The scores are math quiz results from one specific class period

(D) The table shows grades for the whole year

 # End of Practice Test 10 

Great job finishing the test!

 **My Score**

I got _____________ out of 30 questions right.

Check your answers in the **Answer Key** at the back of the book.

💡 Review any questions you missed. That's how we learn!

📊 **Check Your Score Online!**

Visit **ViewMath Academy** to enter your answers and see which topics you need to review. You can also explore lessons, take quizzes, track your scores, and save your progress!

viewmath.com/score/6.1.IN.25

Or go to viewmath.com/score and enter code: 6.1.IN.25

Answer Key & Explanations

Answer Key

First try each test on your own, then check your work here.

Practice Test 1 — Answer Key

1. 5
2. C
3. B
4. D
5. C
6. 216
7. A
8. A
9. -0.625; between -1 and 0
10. 8 blocks
11. B
12. $j + 4 = 13$; $j = 9$
13. $t > 75$
14. B
15. $b = s \div 50$ (or $b = \dfrac{s}{50}$)
16. D
17. 7 cm
18. 4 cm
19. B
20. B
21. B
22. B
23. B
24. B
25. D
26. B
27. B
28. Not always.
29. B
30. positive trend

💡 Time to Learn! 💡

Review the explanations below, **especially for the questions you missed.**

Understanding why each answer is correct builds stronger problem-solving skills.

Tip: Circle any questions you got wrong, then read their explanation carefully.

Practice Test 1 — Detailed Explanations

1. *Each part = $20 \div 4 = 5$. Vegetables = $1 \times 5 = 5$.*

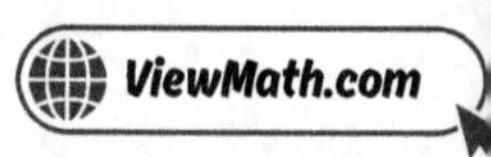

2. From the graph, the truck travels 50 miles in 2 hours. Unit rate: $50 \div 2 = 25$ miles per hour.

3. Ratio: $1 : 4$. When $x = 3$: $y = 3 \times 4 = 12$.

4. Divide by 100: $72 \div 100 = 0.72$.

5. 1 yard $= 3$ feet. $7 \times 3 = 21$ feet.

6. $90 \div 42 = 2$ R6; bring down $7 \to 67 \div 42 = 1$ R25; bring down $2 \to 252 \div 42 = 6$. Answer: 216. Check: $216 \times 42 = 9{,}072$.

7. Point A is at 3.4 and point B is at 5.7. Distance $= 5.7 - 3.4 = 2.3$.

8. Multiples of 4: $4, 8, 12, 16, 20, \ldots$ Multiples of 10: $10, 20, \ldots$ The smallest number in both lists is 20.

9. $5 \div 8 = 0.625$, so $-\dfrac{5}{8} = -0.625$. Since $-1 < -0.625 < 0$, it falls between -1 and 0.

10. Both locations share $y = 4$, so the distance is $|-6 - 2| = |-8| = 8$ blocks.

11. Group 1 has $3x + 1$. Group 2 has $x + 3$. Total: $3x + 1 + x + 3 = 4x + 4$. Choices B and D both represent the same value; B is simplified.

12. Jake's age plus 4 equals 13: $j + 4 = 13$. Subtract 4: $j = 9$. Jake is 9 years old.

13. "Above 75" means strictly greater than 75: $t > 75$.

14. $>$ does not include 4, so use an open circle to show 4 is NOT a solution.

Find more at
ViewMath.com/IN-Grade6

15 Divide total students by 50 per bus: $b = s \div 50$.

16 $A = \frac{1}{2} \times 14 \times 8 = 56 \ m^2$.

17 $84 = \frac{1}{2}(10 + 14)h = \frac{1}{2}(24)h = 12h$. So $h = 84 \div 12 = 7 \ cm$.

18 Base area $= 9 \times 5 = 45$. Height $= 180 \div 45 = 4 \ cm$.

19 A reflection preserves the size and shape. The coordinates change, and the orientation flips, but the triangle stays congruent to the original.

20 Base $= |5 - 1| = 4$. Height $= |9 - 1| = 8$. Area $= \frac{1}{2} \times 4 \times 8 = 16$ square units.

21 Across the x-axis: keep x, change the sign of y. $(-3, -5) \rightarrow (-3, 5)$.

22 Radius $= 16 \div 2 = 8 \ in$. $A = \pi r^2 = 3.14 \times 8^2 = 3.14 \times 64 = 200.96 \ in^2$. Choice A uses πd (circumference). Choice D uses πd^2 instead of πr^2.

23 The answers vary from student to student, which is what makes the question statistical.

24 A peak is where the data has the highest frequency. The value 3 appears three times, more than any other value.

25 Range $= 35 - 5 = 30$.

26 1 hour has 6 dots (highest frequency) ☒ mode. 8 hours is far from the cluster (0–4) and has only 1 dot ☒ outlier.

27 The left whisker extends from the minimum to Q1, covering the lowest 25% of the data.

Find more at
ViewMath.com/IN-Grade6

ViewMath.com

28 A single outlier can make the range very large while the rest of the data is tightly clustered. The IQR gives a better picture of overall spread. For example, $\{1, 50, 51, 52, 53\}$ has range 52 but most values are close together.

29 Numbers ≤ 3: $1, 2, 3$. That is 3 out of 6. $P(\leq 3) = \dfrac{3}{6} = \dfrac{1}{2}$.

30 Since the temperature increases each month from January to May, the line goes upward, which is a positive trend.

✅ Practice Test 2 — Answer Key

1 C

2 Part A: Lily earns $12 per hour; Noah earns $13 per hour.   Part B: Noah earns $1 more per hour.

3 Part A: Runner A $= 3 : 4$; Runner B $= 2 : 3$.   Part B: Runner A is faster.   **4** C   **5** C   **6** A

7 C   **8** GCF $= 4$;   LCM $= 24$ **9** B

10 8 units. Subtract the x-coordinates: $5 - (-3) = 8$. Take the absolute value: $|8| = 8$. The absolute value ensures distance i

11 C   **12** $n - 19 = 31$; $n = 50$ **13** $x \geq 5$

14 Solutions: 0 and 5 (or any values greater than -1). Non-solution: -1 (or any value ≤ -1)   **15** C

16 B   **17** A   **18** C   **19** $(4, 9)$ **20** B   **21** A   **22** 56.52 cm^2 **23** C

24 Center: around 23. Spread: range $= 5$ (not very spread). Shape: roughly symmetric with a peak at 23.

25 C   **26** Total $= 100$, Mode $=$ Bus, Walk $= 15\%$ **27** B   **28** A   **29** C   **30** B

Get Online
Find more at
ViewMath.com/IN-Grade6

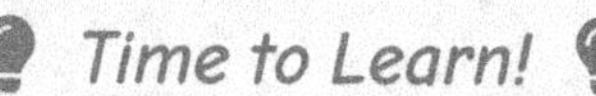

💡 Time to Learn! 💡

Review the explanations below, **especially for the questions you missed**.

Understanding why each answer is correct builds stronger problem-solving skills.

Tip: Circle any questions you got wrong, then read their explanation carefully.

📖 Practice Test 2 — Detailed Explanations

1 Girls have 6 parts $= 24$, so each part $= 24 \div 6 = 4$. Boys have 4 parts $= 4 \times 4 = 16$.

2 Lily: $\$60 \div 5 = \$12$/hr. Noah: $\$52 \div 4 = \$13$/hr. Noah earns $\$13 - \$12 = \$1$ more per hour.

3 Runner A: 3 laps in 4 min $= 0.75$ laps/min. Runner B: 2 laps in 3 min ≈ 0.667 laps/min. $0.75 > 0.667$, so Runner A is faster. On the graph, Runner A's line is steeper.

4 $50\% = \dfrac{50}{100} = \dfrac{1}{2}$.

5 $500 \div 1{,}000 = 0.5$ kg.

6 $1{,}890 \div 15 = 126$. Check: $126 \times 15 = 1{,}890$.

7 Line up the decimals: $20.30 - 8.57$. Regroup as needed: $20.30 - 8.57 = 11.73$.

8 GCF: Factors of 8: $1, 2, 4, 8$. Factors of 12: $1, 2, 3, 4, 6, 12$. Common factors: $1, 2, 4$. GCF $= 4$. LCM: Multiples of 8: $8, 16, 24, \ldots$ Multiples of 12: $12, 24, \ldots$ LCM $= 24$.

9 On a number line, numbers increase from left to right. The correct order from least to greatest is $-1.2 < -0.5 < 0.3 < 1.5$. Choice D shows greatest to least.

Find more at
ViewMath.com/IN-Grade6

10   Since both points share $y = -1$, the distance is $|5 - (-3)| = |5 + 3| = |8| = 8$. Absolute value removes any negative sign, because distance measures how far apart two points are and cannot be negative.

11   Distribute: $6 + 2x + 8$. Combine constants: $6 + 8 = 14$. Result: $2x + 14$.

12   Add 19: $n = 31 + 19 = 50$.

13   Values 5 and above are solutions; values below 5 are not. The inequality is $x \geq 5$.

14   Any number greater than -1 is a solution. -1 itself is not, since $>$ does not include the boundary.

15 $y = 2(4) + 1 = 8 + 1 = 9$.

16 $A = \frac{1}{2} \times 12 \times 9 = 54$ ft^2.

17 $54 = 9 \times h$, so $h = 54 \div 9 = 6$ cm.

18   Each 1-inch cube takes up 1 in^3, so the number of cubes equals the volume: $12 \times 8 \times 6 = 576$.

19   Across the x-axis: change the sign of y. $(4, -9) \rightarrow (4, 9)$.

20   Base $= |8 - 2| = 6$. Height $= |7 - 1| = 6$. Area $= \frac{1}{2} \times 6 \times 6 = 18$ square units.

21   Right 7: $-5 + 7 = 2$. Up 4: $-2 + 4 = 2$. So $K'(2, 2)$.

22   Radius $= 12 \div 2 = 6$ cm. Area of full circle $= \pi r^2 = 3.14 \times 36 = 113.04$ cm^2. Semicircle area $= 113.04 \div 2 = 56.52$ cm^2.

Find more at
ViewMath.com/IN-Grade6

ViewMath.com

23 Exercise time varies from student to student. The other questions each have one fixed answer.

24 The center (median) is 23. Range $= 25 - 20 = 5$. Values are balanced around 23 with the peak there.

25 $IQR = Q3 - Q1 = 55 - 35 = 20.$

26 $45 + 30 + 15 + 10 = 100.$ Bus has the highest frequency ⟹ mode. Walk: $15/100 = 15\%.$

27 The five-number summary consists of minimum, first quartile (Q1), median, third quartile (Q3), and maximum.

28 Same medians, but Athlete A's range (1.2) is smaller — the times vary less, indicating more consistency.

29 Event Y is positioned closest to 0.75 on the probability scale.

30 Boys in Class A: bar reaches 6 units $= 30$ books. Girls in Class A: bar reaches 4 units $= 20$ books. Difference: $30 - 20 = 10.$

✅ Practice Test 3 — Answer Key

 C  B  B  C  C  A  B  A  B  B

 B 12 $a = 7$  C 14 $x \leq -4$  C 16 B 17 D 18 A 19 C

20 B 21 C 22 $C = 43.96$ cm and $A = 153.86$ cm^2 23 C 24 A 25 B

26 Mode $= 5000$, Range $= 4000$ 27 B 28 B 29 C 30 B

Find more at
ViewMath.com/IN-Grade6

🌐 ViewMath.com

> 💡 *Time to Learn!* 💡
>
> *Review the explanations below, **especially for the questions you missed**.*
>
> *Understanding why each answer is correct builds stronger problem-solving skills.*
>
> ***Tip:** Circle any questions you got wrong, then read their explanation carefully.*

📖 Practice Test 3 — Detailed Explanations

1. *The question asks flour to eggs. Flour $= 3$, eggs $= 2$. The ratio is $3 : 2$.*

2. *$\$9 \div 12 = \$0.75$ per muffin.*

3. *The ratio is $6 : 2 = 3 : 1$. When $x = 15$: $y = 15 \div 3 = 5$.*

4. *50% of $200 = \dfrac{1}{2} \times 200 = 100$ items.*

5. *$156 \div 12 = 13$ feet.*

6. *$7 \div 3 = 2$ R1; bring down $5 \to 15 \div 3 = 5$; bring down $6 \to 6 \div 3 = 2$. Answer: 252. Check: $252 \times 3 = 756$.*

7. *Multiply as whole numbers: $12 \times 5 = 60$. Count decimal places: 0.12 has 2 and 0.5 has 1, total 3. Place the decimal: $0.060 = 0.06$.*

8. *Factors of 36: $1, 2, 3, 4, 6, 9, 12, 18, 36$.   Factors of 48: $1, 2, 3, 4, 6, 8, 12, 16, 24, 48$.   Common factors: $1, 2, 3, 4, 6, 12$. The greatest is 12.*

9. *$\dfrac{11}{4} = 2.75$. Since $2 < 2.75 < 3$, it is between 2 and 3.*

Find more at
ViewMath.com/IN-Grade6

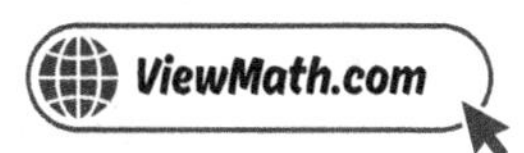

10 Choice A: $|3 - 7| = 4$. Choice B: $|-2 - 4| = 6$. Choice C: $|-3 - 2| = 5$. Choice D: $|-1 - 3| = 4$. The greatest distance is 6 in Choice B.

11 Distribute: $4 \times 3y - 4 \times 2 = 12y - 8$.

12 $a = 105 \div 15 = 7$.

13 "Fewer than 20" means less than 20: $y < 20$.

14 Closed circle means $\leq$ or $\geq$. Shading left means less than or equal to: $x \leq -4$.

15 $i = 12(5) = 60$ inches.

16 $15 \times 4 = 60$ is the rectangle area. Sam forgot to multiply by $\frac{1}{2}$. The correct area is 30 cm^2.

17 $A = \frac{1}{2}(4 + 8)(5) = \frac{1}{2}(12)(5) = 30$ square units.

18 $V = \frac{1}{2} \times \frac{1}{2} \times \frac{1}{2} = \frac{1}{8}$ m^3.

19 Length $= |6 - (-4)| = 10$. Width $= |5 - (-3)| = 8$. Perimeter $= 2(10) + 2(8) = 36$ units.

20 Bottom rectangle: $8 \times 3 = 24$. Left rectangle above: $4 \times (7 - 3) = 4 \times 4 = 16$. Total: 40. This matches option B.

21 Across the y-axis: change the sign of x, keep y. $F(3, 7) \to F'(-3, 7)$.

22 Circumference: $C = \pi d = 3.14 \times 14 = 43.96$ cm. For area, first find $r = 14 \div 2 = 7$ cm. $A = \pi r^2 = 3.14 \times 7^2 = 3.14 \times 49 = 153.86$ cm^2.

Find more at
ViewMath.com/IN-Grade6

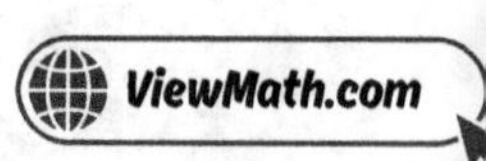

23. *On a single day, the total is one number. However, if she collects data across many days, the number would vary and become statistical. As asked (for one day), it has a single answer.*

24. *The outlier 100 increases the mean significantly. Without it, the center is around 17. With it, the mean jumps to about 27.*

25. *MAD (Mean Absolute Deviation) is calculated by finding the average of the distances of each data value from the mean.*

26. *5000 has frequency 6 (highest). Range $= 7000 - 3000 = 4000$.*

27. *10 values. Median $= (9+11) \div 2 = 10$. Lower half: $1, 3, 5, 7, 9$. $Q1 = 5$. Upper half: $11, 13, 15, 17, 19$. $Q3 = 15$. $IQR = 15 - 5 = 10$.*

28. *A small IQR means the middle 50% is tightly packed. A large range means the minimum and maximum are far apart, indicating extreme values stretch the data.*

29. *There are 6 equally likely outcomes. Only 1 of them is a 4. $P(4) = \dfrac{1}{6}$.*

30. *"At least 2" means 2 or more. Students with 2 pets: 5; with 3 pets: 4. Total: $5 + 4 = 9$.*

☑ Practice Test 4 — Answer Key

1. B   2. 3.5 hours   3. C   4. 87.5%; $\dfrac{7}{8}$ 5. C   6. C   7. C   8. 16 bags

9. A   10. D   11. C   12. $w = 72$ 13. D

14. No. Both shade to the left, but $x < 8$ has an open circle at 8 and $x \leq 8$ has a closed circle at 8.   15. B

16. A   17. B   18. A   19. D   20. B   21. C   22. C   23. B

Find more at
ViewMath.com/IN-Grade6

 Skewed to the left (or: most data on the right with a tail to the left).    C    B    B

 A   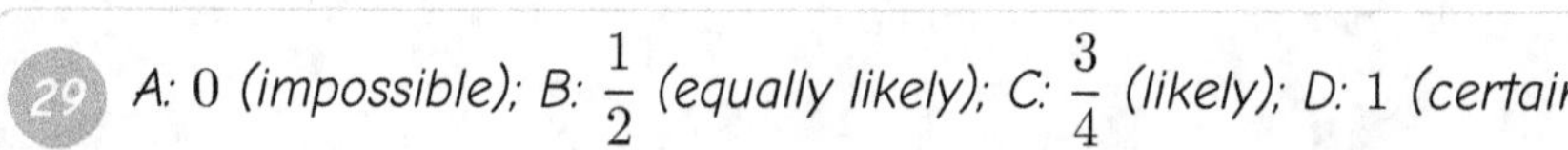 A: 0 (impossible); B: $\frac{1}{2}$ (equally likely); C: $\frac{3}{4}$ (likely); D: 1 (certain)    B

💡 Time to Learn! 💡

Review the explanations below, **especially for the questions you missed.**

Understanding why each answer is correct builds stronger problem-solving skills.

Tip: Circle any questions you got wrong, then read their explanation carefully.

📖 Practice Test 4 — Detailed Explanations

1. "3 pencils for every 1 eraser" means pencils to erasers is $3 : 1$.

2. $49 \div 14 = 3.5$ hours.

3. $5 : 15$ simplifies to $1 : 3$.

4. $0.875 \times 100 = 87.5\%$. $0.875 = \frac{875}{1000} = \frac{7}{8}$.

5. 1 foot = 12 inches. $5 \times 12 = 60$ inches.

6. $15 \div 5 = 3$; bring down $7 \rightarrow 7 \div 5 = 1$ R2; bring down $5 \rightarrow 25 \div 5 = 5$. Answer: 315. Check: $315 \times 5 = 1{,}575$.

7. 0.6 has 1 decimal place and 0.7 has 1 decimal place. Add them: $1 + 1 = 2$ decimal places. The product is 0.42.

Find more at
ViewMath.com/IN-Grade6

ViewMath.com

8 Find the GCF of 48 and 80. Factors of 48: $1, 2, 3, 4, 6, 8, 12, 16, 24, 48$. Factors of 80: $1, 2, 4, 5, 8, 10, 16, 20, 40, 80$. $GCF = 16$. She makes 16 bags, each with 3 pencils and 5 erasers.

9 $\frac{5}{8} = 0.625$. Since $0 < 0.625 < 1$, the fraction lies between 0 and 1.

10 The two points share the same x-coordinate, so the distance is the absolute difference of the y-coordinates: $|-3 - 5| = |-8| = 8$. Choice A forgets the absolute value. Choice B subtracts the x-coordinates. Choice C subtracts the wrong pair.

11 When combining like terms, add coefficients: $2 + 3 = 5$. The variable stays as x, not x^2.

12 Multiply by 8: $w = 9 \times 8 = 72$.

13 "At most" means the value can equal the number or be less, so it should be $\leq$, not $<$.

14 The direction is the same, but the circle type differs: open vs. closed.

15 The independent variable (number of cars) goes on the x-axis.

16 Triangle A: $\frac{1}{2} \times 4 \times 4 = 8$. Triangle B: $\frac{1}{2} \times 5 \times 3 = 7.5$. Triangle A has the greater area.

17 When both bases are equal ($b_1 = b_2 = 4$), the trapezoid is actually a parallelogram. Area $= \frac{1}{2}(4 + 4)(6) = 24 = 4 \times 6$.

18 $V = 1.5 \times 2 \times 4 = 12 \text{ cm}^3$.

19 $|4 - (-2)| = |6| = 6$ units.

20 Length $= |5 - (-1)| = 6$. Width $= |3 - (-2)| = 5$. Area $= 6 \times 5 = 30$ square units.

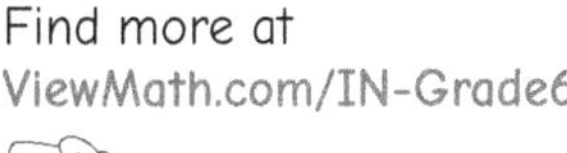
Find more at
ViewMath.com/IN-Grade6

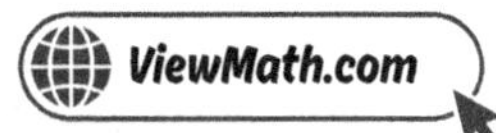

21. Across the y-axis: change the sign of x, keep y. $(2, -6) \to (-2, -6)$.

22. Area of $A = \pi \times 3^2 = 9\pi$. Area of $B = \pi \times 6^2 = 36\pi$. Ratio $= 36\pi \div 9\pi = 4$. When the radius doubles, the area becomes 4 times as large.

23. The dot plot shows data ranging from 0 to 5, with different students giving different answers. This variability confirms the question is statistical.

24. The peak is at 5, with values trailing to the left toward 1. Most data is on the higher end.

25. Upper half: $12, 14, 16$. The median of the upper half is 14, so $Q3 = 14$.

26. Range $= 74 - 68 = 6$.

27. Short whiskers and a narrow box indicate small range and small IQR, meaning data values are close together.

28. $Q_1 = (15 + 18)/2 = 16.5$. $Q_3 = (25 + 28)/2 = 26.5$. $IQR = 26.5 - 16.5 = 10$.

29. Event A: A standard die has faces 1–6, so rolling a 7 is impossible; $P = 0$. Event B: $P(tails) = \dfrac{1}{2}$. Event C: $P(red) = \dfrac{6}{8} = \dfrac{3}{4} = 0.75$. Event D: All faces (1–6) are less than 7, so $P = 1$ (certain).

30. A steep upward line means a sharp increase. A flat line means no change. So the value increased sharply then stayed the same.

✅ Practice Test 5 — Answer Key

 10  C  B  D  C  301  B 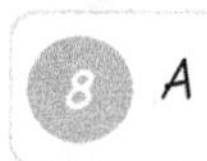 A  B

Find more at
ViewMath.com/IN-Grade6

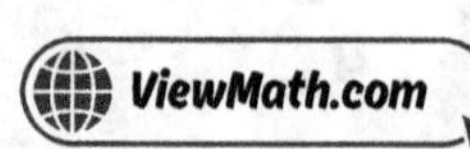

10 $y = 13$ or $y = -3$ **11** $12x - 5$ **12**   C

13   Answers vary. Example: A classroom has at most 15 computers.   **14**  C   **15**  A   **16**  9 m

17   B   **18**   B   **19**   A   **20**   B   **21** $A'(1, -2)$, $B'(4, -2)$, $C'(4, -6)$ **22**   C   **23**   B   **24**  C

25   A   **26**   C   **27**  Range $= 55$, IQR $= 30$ **28**  B   **29**   (a) $\dfrac{3}{8}$ (b) $\dfrac{1}{4}$ (c) $\dfrac{5}{8}$ **30**   B

💡 Time to Learn! 💡

*Review the explanations below, **especially for the questions you missed**.*

Understanding why each answer is correct builds stronger problem-solving skills.

Tip: *Circle any questions you got wrong, then read their explanation carefully.*

📖 Practice Test 5 — Detailed Explanations

1   Total parts $= 3 + 2 = 5$. Each part $= 25 \div 5 = 5$. Swimming $= 2 \times 5 = 10$.

2 $240 \div 8 = 30$ miles per gallon.

3   Ratio $= 3 : 2$. Double: $(6, 4)$. Triple: $(9, 6)$, not $(9, 4)$ or $(9, 8)$.

4   Zoo $= 60\%$. Remaining $= 40\%$. Farm is twice Park, so Farm $= 2x$ and Park $= x$, giving $3x = 40\%$, $x \approx 13.3\%$. Farm $\approx 26.7\% \approx 27\%$.

5 1 km $= 1{,}000$ m. $4.5 \times 1{,}000 = 4{,}500$ m.

6 $45 \div 15 = 3$; bring down $1 \to 1 \div 15 = 0$ R1; bring down $5 \to 15 \div 15 = 1$. Answer: 301. The zero in the tens place is important. Check: $301 \times 15 = 4{,}515$.

Find more at
ViewMath.com/IN-Grade6

7   Multiply as whole numbers: $12 \times 3 = 36$. Count decimal places: 1.2 has 1 and 0.3 has 1, total 2. Place the decimal: 0.36.

8   Find the GCF of 72 and 96.   Factors of 72: $1, 2, 3, 4, 6, 8, 9, 12, 18, 24, 36, 72$.   Factors of 96: $1, 2, 3, 4, 6, 8, 12, 16, 24, 32, 48, 96$. Common factors include $1, 2, 3, 4, 6, 8, 12, 24$. The greatest is 24 inches.

9 $-\dfrac{7}{3} \approx -2.333$. Since $-3 < -2.333 < -2$, it is between -3 and -2.

10   The points share $x = -2$, so the distance is $|y - 5| = 8$. This means $y - 5 = 8$ or $y - 5 = -8$. Solving: $y = 13$ or $y = -3$. Both values give a distance of 8.

11 $9x + 3x = 12x$ and $2 - 7 = -5$. Result: $12x - 5$.

12   Multiply both sides by 4: $m = 7 \times 4 = 28$.

13   Any situation where a quantity is 15 or fewer works.

14 $-3 \geq -3$ is true (they are equal). The other values are less than -3.

15 $y = 1 + 4 = 5$, $y = 2 + 4 = 6$, $y = 3 + 4 = 7$. Match: A.

16 $45 = \frac{1}{2} \times 10 \times h$, so $45 = 5h$, giving $h = 9$ m.

17 $A = \frac{1}{2}(9 + 15)(6) = \frac{1}{2}(24)(6) = 72$ in^2.

18 $V = 8 \times 6 \times 3 = 144$ in^3.

19   Across the x-axis, change the sign of each y-coordinate: $(1, -3)$, $(5, -3)$, $(5, -7)$.

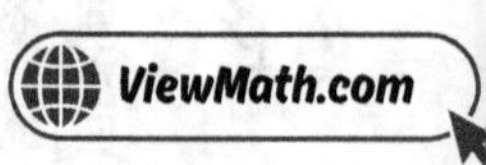

20   *Width* $= 56 \div 8 = 7$ *units.*

21   *Across the x-axis: change the sign of each y-coordinate.* $(1,2) \to (1,-2)$, $(4,2) \to (4,-2)$, $(4,6) \to (4,-6)$.

22 $A = \pi r^2 = 3.14 \times 5^2 = 3.14 \times 25 = 78.5$ *in*^{2}. *Choice B uses the circumference formula* $2\pi r$. *Choice D uses* $2\pi r^2$.

23   *Tree heights vary from tree to tree, and running speeds vary from student to student. Both questions expect varying data.*

24   *Range* $= 18 - 12 = 6$. *The range tells you how spread out the data is.*

25   *The mean is* 10. *Every value equals* 10, *so every distance from the mean is* 0. *MAD* $= 0 \div 5 = 0$.

26   *Since two values share the highest frequency of* 4, *the data set is bimodal — it has* 2 *modes.*

27   *Range* $= 80 - 25 = 55$. *IQR* $= 65 - 35 = 30$.

28   *Same median (same typical sales). Store A's range ($200) is much smaller than Store B's ($800), making Store A more consistent.*

29   *There are 8 equal sections: 3 red, 2 blue, 3 green. (a) P(red)* $= \dfrac{3}{8}$. *(b) P(blue)* $= \dfrac{2}{8} = \dfrac{1}{4}$. *(c) P(not green)* $= 1 - \dfrac{3}{8} = \dfrac{5}{8}$.

30   *The axes show sports (categories) and number of students (counts). The best title describes a survey of students' favorite sports.*

✅ Practice Test 6 — Answer Key

Find more at
ViewMath.com/IN-Grade6

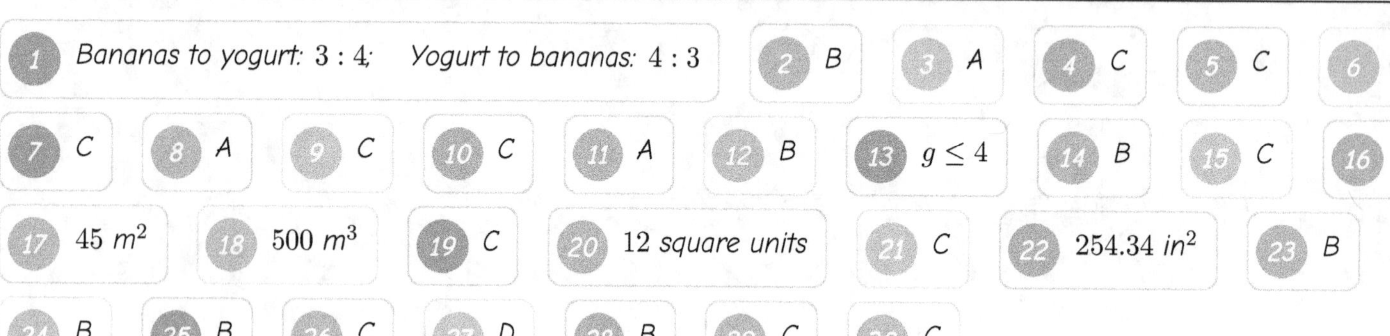

1	Bananas to yogurt: $3 : 4$; Yogurt to bananas: $4 : 3$
2	B
3	A
4	C
5	C
6	B
7	C
8	A
9	C
10	C
11	A
12	B
13	$g \leq 4$
14	B
15	C
16	C
17	$45\ m^2$
18	$500\ m^3$
19	C
20	12 square units
21	C
22	$254.34\ in^2$
23	B
24	B
25	B
26	C
27	D
28	B
29	C
30	C

💡 Time to Learn! 💡

Review the explanations below, **especially for the questions you missed**.

Understanding why each answer is correct builds stronger problem-solving skills.

Tip: *Circle any questions you got wrong, then read their explanation carefully.*

📖 Practice Test 6 — Detailed Explanations

1. *"3 bananas for every 4 cups of yogurt" gives $3 : 4$. Flip the order for yogurt to bananas: $4 : 3$.*

2. *Brand X: $\$8.75 \div 5 = \$1.75$ each. Brand Y: $\$4.50 \div 3 = \$1.50$ each. Brand Y is cheaper.*

3. *The ratio is $2 : 3$. Choice A continues with $6 : 9 = 2 : 3$. Choice B has $6 : 8$ which is not $2 : 3$.*

4. *Multiply by 100: $0.35 \times 100 = 35\%$.*

5. *$1\ km = 1{,}000\ m = 100{,}000\ cm$. $10 \times 100{,}000 = 1{,}000{,}000\ cm$.*

6. *$4{,}200 \div 7 = 600$. Estimating with compatible numbers makes division easier.*

Find more at
ViewMath.com/IN-Grade6

7 Move the decimal one place in both numbers: $126 \div 9 = 14$. Check: $14 \times 0.9 = 12.6$.

8 Multiples of 3: $3, 6, 9, \ldots$ Multiples of 9: $9, 18, \ldots$ The smallest number in both lists is 9. When one number is a multiple of the other, the LCM is the larger number.

9 $-\dfrac{3}{4} = -0.75$. Since $-1 < -0.75 < 0$, it lies between -1 and 0. To plot it, divide the segment from -1 to 0 into 4 equal parts and count 3 parts to the left of 0.

10 The points share $x = 1$, so the distance is $|-2 - 5| = |-7| = 7$ units. Choice A subtracts the y-values incorrectly as $5 - 2$. Choice B forgets the absolute value. Choice D adds the absolute values $|-2| + |5| + |1| + |1|$.

11 Distribute: $7k - 21 + 5$. Combine: $-21 + 5 = -16$. Result: $7k - 16$.

12 Left: $x + 3$. Right: 10. Equation: $x + 3 = 10$. Subtract 3: $x = 7$.

13 "No more than 4" means 4 or fewer. $g \leq 4$.

14 "More than 75" is $t > 75$: open circle (not including 75), shade right (greater values).

15 Each dog has 4 legs: $L = 4d$.

16 $A = \frac{1}{2} \times 2.4 \times 5 = 6 \ m^2$.

17 $A = \frac{1}{2}(7 + 11)(5) = \frac{1}{2}(18)(5) = 45 \ m^2$.

18 $V = 25 \times 10 \times 2 = 500 \ m^3$.

19 $|5 - (-3)| = |8| = 8 \ units$.

Find more at
ViewMath.com/IN-Grade6

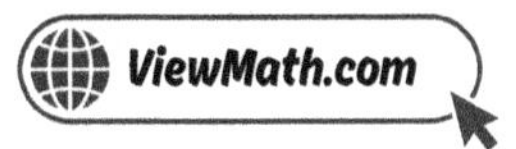

20 Rectangle area $= 6 \times 4 = 24$. Triangle area $= \frac{1}{2} \times 6 \times 4 = 12$. Remaining $= 24 - 12 = 12$ square units.

21 The y-coordinate stayed the same. The x-coordinate changed from -2 to 4: $4 - (-2) = 6$ units right. Note: reflection across the y-axis would give $(2, 7)$, not $(4, 7)$.

22 First find the radius: $r = 18 \div 2 = 9$ in. Then $A = \pi r^2 = 3.14 \times 9^2 = 3.14 \times 81 = 254.34$ in^2.

23 Different students drink different amounts, so the answers vary. The other choices still have single fixed answers.

24 Most scores cluster around 70–76. The center is about 73–75. The 98 is an outlier that would pull the mean higher, but the typical values are in the low-to-mid 70s.

25 Distances from 24: $4, 2, 0, 2, 4$. $MAD = (4 + 2 + 0 + 2 + 4) \div 5 = 12 \div 5 = 2.4$.

26 Favorite color is categorical, not numerical. A frequency table or bar graph handles categories; histograms and dot plots need numerical data.

27 Different data sets can have the same five-number summary. The five-number summary only captures five key values, not every data point.

28 Week 2 median (9,500) is higher. Week 2 IQR (1,200) is smaller, meaning more consistent daily step counts.

29 A fair coin has 2 equally likely outcomes (heads or tails). $P(heads) = \dfrac{1}{2}$.

30 Add all frequencies: $5 + 8 + 3 + 4 = 20$ students.

✅ Practice Test 7 — Answer Key

Find more at
ViewMath.com/IN-Grade6

1   6   **2**   B   **3** $(2, 5)$ and $(8, 20)$ *(or any equivalent pair)*   **4**   B

5   1.5 *quarts; No, one 1-quart container is not enough.*   **6**   B   **7**   2.4   **8**   C   **9**   C

10   C   **11** $5(3n + 2)$ **12**   C   **13**   B   **14**   D   **15** $c = 5m + 10$; *after 6 months:* $40

16   C   **17**   B   **18**   A   **19**   C   **20**   64 *square units*   **21**   A   **22**   B   **23**   *Statistical*

24   D   **25**   20   **26**   C

27   *Store B has higher typical sales (median $\approx$ $275 vs. $250). Store A has more predictable sales (IQR $\approx$ $100 vs. $125).*

28   B   **29**   B   **30**   B

💡 Time to Learn! 💡

*Review the explanations below, **especially for the questions you missed**.*

Understanding why each answer is correct builds stronger problem-solving skills.

***Tip:** Circle any questions you got wrong, then read their explanation carefully.*

📖 Practice Test 7 — Detailed Explanations

1 $21 \div 7 = 3$, *so multiply both by 3: iced tea* $= 2 \times 3 = 6$.

2 $150 \div 5 = 30$ *gallons per hour.*

3   *Ratio* $= 4 : 10 = 2 : 5$. *Half gives* $(2, 5)$; *double gives* $(8, 20)$.

4   *70% of* $30 = 0.70 \times 30 = 21$ *students.*

5 $3 \div 2 = 1.5$ quarts. Since $1.5 > 1$, one 1-quart container is not enough.

6 Divide step by step: $9 \div 4 = 2$ R1; bring down $3 \to 13 \div 4 = 3$ R1; bring down $6 \to 16 \div 4 = 4$. Answer: 234. Check: $234 \times 4 = 936$.

7 Move the decimal two places in both numbers: $28.8 \div 12 = 2.4$. Check: $2.4 \times 0.12 = 0.288$.

8 Factors of 60: $1, 2, 3, 4, 5, 6, 10, 12, 15, 20, 30, 60$. Factors of 84: $1, 2, 3, 4, 6, 7, 12, 14, 21, 28, 42, 84$. Common factors: $1, 2, 3, 4, 6, 12$. The greatest is 12.

9 A rational number can be written as $\frac{a}{b}$ where $b \neq 0$. The fraction $-\frac{3}{4}$ fits this definition. The numbers $\sqrt{2}$, π, and $\sqrt{5}$ are irrational because they cannot be expressed as a fraction of two integers.

10 The lights share $y = 5$, so the distance is $|-3 - 4| = |-7| = 7$ blocks. Choice B uses only the y-coordinate. Choice D adds $|-3| + |4| + 2$ or a similar error.

11 GCF of 15 and 10 is 5. Factor: $5(3n + 2)$.

12 $3(5) = 15 \neq 18$. So $x = 5$ is not a solution. The correct solution is $x = 6$.

13 "Over 13" means greater than 13: $a > 13$.

14 $\leq$ includes -1 (closed circle). Less than -1 is to the left (shade left).

15 $c = 5(6) + 10 = 30 + 10 = 40$ dollars.

16 $A = \frac{1}{2} \times 8 \times 5 = 20$ in^2.

17 $A = b \times h = 9 \times 5 = 45$ cm^2.

Find more at
ViewMath.com/IN-Grade6

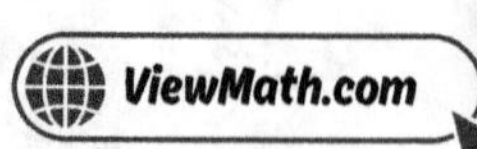

18 $V = \frac{3}{4} \times 2 \times 4 = \frac{3}{4} \times 8 = 6 \ ft^3$.

19 Across the y-axis: change the sign of x. $(-6, -4) \rightarrow (6, -4)$.

20 Bottom rectangle: $10 \times 4 = 40$. Top-left rectangle: $6 \times (8 - 4) = 6 \times 4 = 24$. Total: $40 + 24 = 64$ square units.

21 Translate 3 units left: subtract 3 from x. $B(6, 1) \rightarrow B'(6 - 3, 1) = (3, 1)$.

22 One full rotation covers the circumference. $C = 2\pi r = 2 \times 3.14 \times 13 = 81.64$ in. Choice A forgets the factor of 2. Choice D uses the area formula.

23 Different students get different amounts of sleep, so the answers would vary from student to student.

24 Most values are between 2 and 7, clustered around 4. The value 20 is far away from the rest, making it an outlier.

25 Median $= 115$. Q1: median of $100, 105, 110 = 105$. Q3: median of $120, 125, 130 = 125$. IQR $= 125 - 105 = 20$.

26 Every value is 5, so 5 is the value that appears most often. The mode is 5.

27 Store A: median $\approx \$250$, IQR $\approx 300 - 200 = \$100$. Store B: median $\approx \$275$, IQR $\approx 350 - 225 = \$125$. Store B sells more on a typical day, but Store A's sales are more consistent (smaller IQR and range).

28 The median describes a typical value (center) and the IQR describes how spread out the middle 50% of values are (spread). Together they summarize the data well.

29 $\frac{2}{5} = 0.4$ and $\frac{3}{4} = 0.75$. Since $0.75 > 0.4$, Event B is more likely.

30 Total: $24 + 30 = 54$. Fraction of boys: $\frac{24}{54} = \frac{4}{9}$.

Find more at
ViewMath.com/IN-Grade6

✅ Practice Test 8 — Answer Key

1 B

2 Part A: 0.5 laps per minute (or 1 lap every 2 minutes).   Part B: 7 laps.

3 False. Equivalent ratios always form a straight line through the origin.   **4** C

5 Yes. 3 tons = 6,000 pounds, which is more than 5,500.   **6** A   **7** B   **8** B   **9** B

10 Part A: 7 units;   Part B: 7 units;   Part C: 14 units   **11** B   **12** C

13 6, 7, and 10 (or any three numbers that are 6 or greater)   **14** Graph A: $x < 2$. Graph B: $x \geq -1$.

15 $w = 45 - 3t$; empty when $t = 15$ minutes   **16** $\frac{1}{4}$ ft^2 **17** 12 ft   **18** B   **19** $(-2, 4)$ **20** A

21 $(3, 3)$ **22** B   **23** C   **24** 50 is an outlier. It pulls the center (mean) higher than the typical values.

25 6   **26** B   **27** 50%   **28** C   **29** 20%   **30** D

💡 Time to Learn! 💡

Review the explanations below, **especially for the questions you missed.**

Understanding why each answer is correct builds stronger problem-solving skills.

Tip: Circle any questions you got wrong, then read their explanation carefully.

📖 Practice Test 8 — Detailed Explanations

1 Dogs have 4 parts. Each part = 2 animals. Dogs = 4 × 2 = 8.

2 Part A: From the graph, the swimmer completes 1 lap every 2 minutes, so the rate is $\frac{1}{2}$ lap per minute. Part B: $14 \div 2 = 7$ laps.

3 When you multiply both parts of a ratio by the same number, the points increase at a constant rate, which produces a straight line through $(0,0)$.

4 $100\% = \frac{100}{100} = 1$ whole.

5 $3 \times 2,000 = 6,000$ lb. $6,000 > 5,500$, so yes.

6 Apples: $2,688 \div 16 = 168$. Oranges: $1,575 \div 25 = 63$. Bananas: $1,890 \div 15 = 126$. Apples need the most crates (168).

7 $24 \times 3 = 72$. The correct product has 2 decimal places: 0.72. The student placed only 1 decimal place and got 7.2.

8 $36 \div 8 = 4.5$, so 8 does not divide evenly into 36. It is a factor of 24 but not of 36, so it is not a common factor. The other choices (4, 6, 12) all divide evenly into both 24 and 36.

9 Find the distance from 0: $|0.8| = 0.8$, $\left|-\frac{3}{2}\right| = 1.5$, $|1.1| = 1.1$, $\left|-\frac{4}{5}\right| = 0.8$. The greatest distance is 1.5, so $-\frac{3}{2}$ is farthest from 0.

10 Part A: Same $y = 3$, distance $= |-5 - 2| = |-7| = 7$. Part B: Same $x = 2$, distance $= |3 - (-4)| = |7| = 7$. Part C: Total $= 7 + 7 = 14$ units.

11 The table confirms the same output for every input, and the distributive property proves equivalence: $2(x + 3) = 2x + 6$.

12 Multiply by 6: $t = 5 \times 6 = 30$.

Find more at
ViewMath.com/IN-Grade6

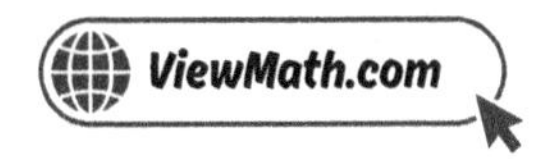

13. Any number greater than or equal to 6 is a solution.

14. Graph A: open circle at 2, shade left $= x < 2$. Graph B: closed circle at -1, shade right $= x \geq -1$.

15. Set $w = 0$: $0 = 45 - 3t$, so $3t = 45$, $t = 15$ minutes.

16. $A = \frac{1}{2} \times \frac{3}{4} \times \frac{2}{3} = \frac{1}{2} \times \frac{6}{12} = \frac{1}{2} \times \frac{1}{2} = \frac{1}{4}$ ft^2.

17. $96 = b \times 8$, so $b = 96 \div 8 = 12$ ft.

18. $V = 3 \times 2 \times 2 = 12$ ft^3.

19. Across the y-axis: change the sign of x. $(2, 4) \to (-2, 4)$.

20. Bottom rectangle: $8 \times 3 = 24$. Top-left rectangle: $4 \times (7 - 3) = 4 \times 4 = 16$. Total: $24 + 16 = 40$ square units.

21. Reflect across the y-axis: $(-3, 5) \to (3, 5)$. Translate 2 units down: $(3, 5 - 2) = (3, 3)$.

22. $C = \pi d = 3.14 \times 10 = 31.4$ cm. Choice A uses πr incorrectly as the full circumference, and choice C doubles the correct answer.

23. A statistical question is one where you expect different answers from different people or situations.

24. Most values are between 10 and 18. The outlier 50 raises the mean from about 14.2 (without it) to about 19.3 (with it).

25. Distances from 60: $10, 5, 0, 5, 10$. $MAD = (10 + 5 + 0 + 5 + 10) \div 5 = 30 \div 5 = 6$.

26 Favorite fruit is categorical data (not numerical intervals). A frequency table counts each category and is the natural first step.

27 Q1 to Q3 is the box, which always contains 50% of the data.

28 The median and IQR are resistant to outliers. The mean and range can be greatly affected by extreme values.

29 Total marbles: $4 + 6 + 10 = 20$. $P(green) = \dfrac{4}{20} = \dfrac{1}{5} = 0.20 = 20\%$.

30 Red: 12, Yellow: 4. Difference: $12 - 4 = 8$.

✓ Practice Test 9 — Answer Key

1 C   2 B   3 A   4 C   5 B   6 C   7 B

8 Part A: $1, 2, 3, 4, 6, 8, 12, 24$;   Part B: 24;   Part C: 24 boxes, each with 2 cookies and 3 brownies   9 C

10 Part A: 9 units;   Part B: 5 units;   Part C: 14 units   11 A   12 B   13 C   14 B   15 B

16 C   17 C   18 A   19 D   20 A   21 D   22 B   23 B   24 C   25 B

26 B   27 A   28 B   29 B   30 Monday, 0 more (they are equal at 35)

💡 Time to Learn! 💡

Review the explanations below, **especially for the questions you missed**.

Understanding why each answer is correct builds stronger problem-solving skills.

Tip: Circle any questions you got wrong, then read their explanation carefully.

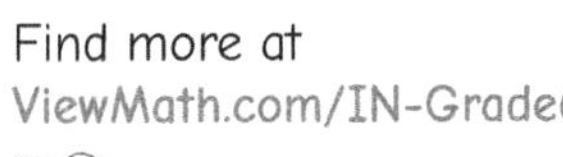
Find more at
ViewMath.com/IN-Grade6

📖 *Practice Test 9 — Detailed Explanations*

1. Total parts $= 1 + 3 + 2 = 6$. Each part $= 18 \div 6 = 3$. Blue $= 3 \times 3 = 9$.

2. Divide: $120 \div 4 = 30$ pages per minute.

3. Line A rises 4 for every 1 it moves right; Line B rises 3. A higher rise means steeper.

4. 88 out of $100 = 88\%$. When the total is 100, the number correct IS the percent.

5. $3 \times 8 = 24$ pints.

6. $1{,}260 \div 18 = 70$. Estimate: $1{,}200 \div 20 = 60$, close. Check: $70 \times 18 = 1{,}260$.

7. Each day she walks $1.25 \times 2 = 2.50$ miles. In 5 days: $2.50 \times 5 = 12.5$ miles.

8. Part A: Numbers appearing in both rows: $1, 2, 3, 4, 6, 8, 12, 24$. Part B: The greatest of these common factors is 24. Part C: $48 \div 24 = 2$ cookies per box and $72 \div 24 = 3$ brownies per box.

9. Find the distance from 0: $\left|-\frac{2}{3}\right| \approx 0.667$, $\left|\frac{3}{4}\right| = 0.75$, $\left|-\frac{1}{5}\right| = 0.2$, $\left|\frac{5}{6}\right| \approx 0.833$. The smallest distance is 0.2, so $-\frac{1}{5}$ is closest to 0.

10. Part A: $A(-4, 3)$ and $B(5, 3)$ share $y = 3$. Distance $= |-4 - 5| = |-9| = 9$. Part B: $B(5, 3)$ and $C(5, -2)$ share $x = 5$. Distance $= |3 - (-2)| = |5| = 5$. Part C: Total $= 9 + 5 = 14$ units.

11. $9w - 4w = 5w$ and $6 - 6 = 0$. Result: $5w$.

Find more at
ViewMath.com/IN-Grade6

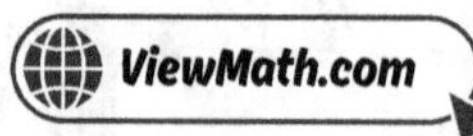

12   Subtract 9 from both sides: $x = 15 - 9 = 6$.

13   "More than \$25" means greater than 25: $d > 25$.

14 $\leq$ means less than or equal to 6. Numbers less than 6 are to the left, so shade left.

15 $y = 4(3) = 12$. The point $(3, 12)$ satisfies $y = 4x$.

16   Rectangle area $= 6 \times 10 = 60$ cm^2. Triangle area $= \frac{1}{2} \times 6 \times 10 = 30$ cm^2. The triangle is half.

17   Both use $A = b \times h = 7 \times 4 = 28$ m^2. They have equal area.

18   Base area $= 10 \times 4 = 40$. Height $= 120 \div 40 = 3$ cm.

19 $(2, 5)$ reflects to $(2, -5)$. Distance $= |5 - (-5)| = 10$ units.

20   Rectangle: $5 \times 4 = 20$. Triangle: $\frac{1}{2} \times 5 \times 3 = 7.5$. Total: $20 + 7.5 = 27.5$ square units.

21   A translation slides every point the same distance in the same direction. Size, shape, and orientation are all preserved.

22   Pi (π) is defined as the ratio of a circle's circumference to its diameter. For every circle, $\pi = \dfrac{C}{d} \approx 3.14$.

23   A dot plot needs data that varies. Only the question about books read would produce different values from different students.

24   An outlier is a value that is far from the rest of the data set.

25. Data Set A: range $= 9 - 5 = 4$. Data Set B: range $= 10 - 4 = 6$. Data Set B has the larger range.

26. The size of the data set and whether it is numerical or categorical determine the best display. Categorical ⊠ frequency table/bar graph. Small numerical ⊠ dot plot. Large numerical ⊠ histogram.

27. Class A's box (IQR) is narrower than Class B's. A narrower box means the middle 50% of scores are closer together — more consistent.

28. The range uses only the maximum and minimum. Changing any extreme value changes the range. However, the IQR only uses Q_1 and Q_3 and can stay the same even if extremes change. Still, the range is most directly influenced by outliers since it depends on the two most extreme values.

29. $P(cherry) = \dfrac{4}{10}$ and $P(grape) = \dfrac{4}{10}$. The probabilities are equal, so these events are equally likely.

30. Monday: $15 + 20 = 35$. Tuesday: $25 + 10 = 35$. Both days had the same total of 35 fruits.

✅ Practice Test 10 — Answer Key

1. C	2. B	3. C
4. A	5. B	6. 1
7. 15.72	8. 28 seconds	9. B
10. C	11. Error in Step 2. Should be $6x + 12 + x$. Correct answer: $7x + 12$	12. C
13. $p \le 12$ and $w \le 2{,}000$	14. C	15. B
16. 88 cm^2	17. A	18. 84 cubic units
19. B	20. B	21. B
22. B	23. Statistical	24. B
25. C	26. 11	27. B
28. C	29. C	30. C

Find more at
ViewMath.com/IN-Grade6

💡 Time to Learn! 💡

Review the explanations below, **especially for the questions you missed**.

Understanding why each answer is correct builds stronger problem-solving skills.

Tip: Circle any questions you got wrong, then read their explanation carefully.

📖 Practice Test 10 — Detailed Explanations

1. Apple has 3 parts, orange has 5 parts. Total parts $= 8$. Total ounces $= 8 \times 4 = 32$.

2. Store A: $\$4.50 \div 10 = \$0.45$. Store B: $\$3.20 \div 8 = \$0.40$. Store B is cheaper per pencil.

3. At $\$2$ per bottle, 4 bottles cost $\$8$, not $\$10$. The point $(4, 10)$ does not fit the ratio.

4. Convert all to decimals: $30\% = 0.30$, $\frac{1}{3} \approx 0.333$, 0.35. Order: $0.30 < 0.333 < 0.35$.

5. Converting 5 feet to inches uses the conversion ratio $\frac{12\,in}{1\,ft}$.

6. After subtracting 15, the remainder is 0. Bring down 7 to get 7. Since $7 \div 5 = 1$ R2, the missing digit in the quotient is 1. The full answer is 315.

7. Line up the decimals: $25.10 - 9.38$. Hundredths: $0 - 8$, regroup $\rightarrow 10 - 8 = 2$. Tenths: $0 - 3$, regroup $\rightarrow 10 - 3 = 7$. Ones: $4 - 9$, regroup $\rightarrow 14 - 9 = 5$. Tens: 1. Answer: 15.72.

8. Find the LCM of 4 and 7. Multiples of 4: $4, 8, 12, 16, 20, 24, 28, \ldots$ Multiples of 7: $7, 14, 21, 28, \ldots$ LCM $= 28$. They blink together again after 28 seconds.

Find more at
ViewMath.com/IN-Grade6

🌐 ViewMath.com

9 $-\dfrac{5}{3} \approx -1.667$. Since $-2 < -1.667 < -1$, it is between -2 and -1. A common mistake is placing it between -1 and 0 by confusing $\dfrac{5}{3}$ with $\dfrac{2}{3}$.

10 The points share the same x-coordinate, so the distance is $|-6 - (-1)| = |-6 + 1| = |-5| = 5$. Choice B forgets the absolute value. Choice D adds the absolute values $6 + 1$ instead of subtracting.

11 The student distributed 3 to $2x$ but not to 4. Correct: $3 \times 4 = 12$. So $6x + 12 + x = 7x + 12$.

12 Divide by 12: $p = 84 \div 12 = 7$.

13 "Maximum 12 persons" means $p \leq 12$. "Maximum weight 2,000 lbs" means $w \leq 2,000$.

14 $0 > -5$ is true. Since 0 is in the shaded region (right of -5), it confirms the graph.

15 In B, y is always 7 no matter what x is. The value of y does not change with x.

16 $A = \frac{1}{2} \times 16 \times 11 = 88$ cm^2.

17 $11 \times 3 = 33$ cm^2. Maria got 16.5 because she divided by 2, using the triangle formula by mistake.

18 Bottom prism: $8 \times 3 \times 2 = 48$. Top prism: $3 \times 3 \times 4 = 36$. Total: $48 + 36 = 84$ cubic units.

19 A reflected point is always the same distance from the axis of reflection, just on the opposite side. Size does not change.

20 Decompose irregular shapes into simpler shapes (rectangles and triangles) whose areas you can compute, then add them.

21 Reflecting across the x-axis flips the point vertically, so the y-coordinate changes sign: $(x, y) \rightarrow (x, -y)$.

Find more at
ViewMath.com/IN-Grade6

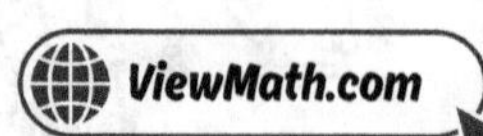

22 $C = \pi d = 3.14 \times 8 = 25.12$ ft. Choice A uses πr without doubling. Choice C confuses with the area formula $\pi r^2 = 3.14 \times 16 = 50.24$.

23 Different students send different numbers of texts, so the answers vary.

24 Data Set A has bars that are tallest in the middle and shorter on the sides (symmetric). Data Set B has the tallest bar on the left with bars decreasing to the right (skewed right).

25 Adding 5 to every value shifts the max and min by the same amount, so the difference (range) stays unchanged.

26 0–14: 3 people. 15–29: 8 people. Total < 30 min: $3 + 8 = 11$.

27 The box stretches from Q1 to Q3, containing the middle 50% of the data.

28 A: $IQR = 30 - 15 = 15$. B: $IQR = 35 - 20 = 15$. Both have $IQR = 15$.

29 Total balls: $5 + 3 + 2 = 10$. $P(yellow) = \dfrac{2}{10}$. $P(not\ yellow) = 1 - \dfrac{2}{10} = \dfrac{8}{10}$.

30 The title tells us the data is about math quiz scores and specifically from Period 3, not the whole school or year.

Well done checking your answers!

Keep practicing to strengthen your skills.

Find more at
ViewMath.com/IN-Grade6

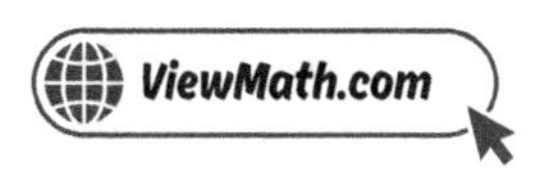

Author's Final Note

I hope you enjoyed this book as much as I enjoyed writing it. Whether you are a student working through the materi
a parent supporting your child's learning, or a teacher guiding your class, I have tried to make this book as clear and
engaging as possible. I hope I have succeeded. If you have any suggestions for improvement, please let me know. I
would love to hear from you.

The accuracy of calculations is very important to me. We have done our best, but I also expect that I have made som
minor errors. Constant improvement is the name of the game. If you find any errors, please let me know. I will fix
them in the next edition.

For students: Your learning journey does not end here. I have written a series of books to help you learn math. Make
sure you browse through them. I especially recommend workbooks and practice tests to help you prepare for your
exams.

For parents: Thank you for investing in your child's education. I encourage you to explore the companion resources
available online to help support your child outside the classroom.

For teachers: Thank you for the invaluable work you do every day. I hope this book serves as a useful resource in you
classroom. Feel free to reach out if you have suggestions or would like to discuss how best to use this book with your
students.

I also enjoy reading your reviews. If you have a moment, please leave a review on where you found this book. It will
help others find this book. If you have any questions or comments, please feel free to contact me at
drNazari@ViewMath.com.

And one last thing: Remember to use online resources for additional help. I recommend using the resources on
`https://ViewMath.com` You can find video lessons, practice problems, and more. You can also use the online
companion for this book to track your progress and access additional resources.

Wishing all students the best in their studies, parents every success in supporting their children, and teachers
continued inspiration in their classrooms!

Dr. A. Nazari

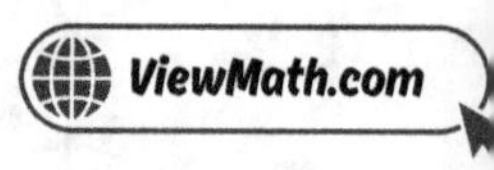

📖 Great Job! Keep Learning with ViewMath!

*Keep up the great work! Visit **viewmath.com/IN-Grade6** for free lessons, quizzes, and more.*

Study Guide

Scan Me

Workbook

Scan Me

Step-by-Step

Scan Me

3 Practice Tests

Scan Me

5 Practice Tests

Scan Me

7 Practice Tests

Scan Me

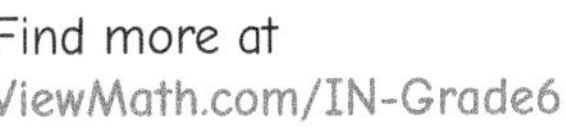
Find more at
ViewMath.com/IN-Grade6

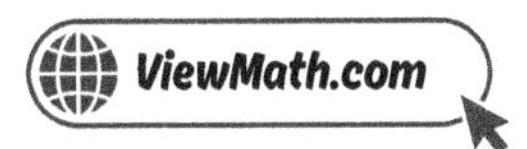